PERCEPT, DECISION, ACTION: BRIDGING THE GAPS

The Novartis Foundation is an international scientific and educational charity (UK Registered Charity No. 313574). Known until September 1997 as the Ciba Foundation, it was established in 1947 by the CIBA company of Basle, which merged with Sandoz in 1996, to form Novartis. The Foundation operates independently in London under English trust law. It was formally opened on 22 June 1949.

The Foundation promotes the study and general knowledge of science and in particular encourages international co-operation in scientific research. To this end, it organizes internationally acclaimed meetings (typically eight symposia and allied open meetings and 15–20 discussion meetings each year) and publishes eight books per year featuring the presented papers and discussions from the symposia. Although primarily an operational rather than a grant-making foundation, it awards bursaries to young scientists to attend the symposia and afterwards work with one of the other participants.

The Foundation's headquarters at 41 Portland Place, London W1B 1BN, provide library facilities, open to graduates in science and allied disciplines. Media relations are fostered by regular press conferences and by articles prepared by the Foundation's Science Writer in Residence. The Foundation offers accommodation and meeting facilities to visiting scientists and their societies.

Information on all Foundation activities can be found at http://www.novartisfound.org.uk

Novartis Foundation Symposium 270

PERCEPT, DECISION, ACTION:
BRIDGING THE GAPS

2006

John Wiley & Sons, Ltd

Other Wiley Editorial Offices

John Wiley & Sons Inc., 111 River Street, Hoboken, NJ 07030, USA

Jossey-Bass, 989 Market Street, San Francisco, CA 94103-1741, USA

Wiley-VCH Verlag GmbH, Boschstr. 12, D-69469 Weinheim, Germany

John Wiley & Sons Australia Ltd, 33 Park Road, Milton, Queensland 4064, Australia

John Wiley & Sons (Asia) Pte Ltd, 2 Clementi Loop #02-01, Jin Xing Distripark, Singapore
129809

John Wiley & Sons Canada Ltd, 22 Worcester Road, Etobicoke, Ontario, Canada M9W 1L1

Wiley also publishes its books in a variety of electronic formats. Some content that appears
in print may not be available in electronic books.

Novartis Foundation Symposium 270
xii + 301 pages, 65 figures, 0 tables

British Library Cataloguing in Publication Data

A catalogue record for this book is available from the British Library

ISBN-13 978-0-470-01233-8
ISBN-10 0-470-01233-1

Typeset in $10\frac{1}{2}$ on $12\frac{1}{2}$ pt Garamond by SNP Best-set Typesetter Ltd., Hong Kong.
Printed and bound in Great Britain by T. J. International Ltd, Padstow, Cornwall.
This book is printed on acid-free paper responsibly manufactured from sustainable forestry,
in which at least two trees are planted for each one used for paper production.

Contents

Jacques Mehler, Marina Nespor, Mohinish Shukla and **Marcela Peña**

Participants

Thomas D. Albright Visual Center Laboratory, The Salk Institute, 10010 N. Torrey Pines Road, La Jolla, CA 92037, USA

Shabtai Barash Department of Neurobiology, Weizmann Institute of Science, Rehovot, 76100, Israel

Michael Brecht Erasmus MC, University Medical Center Rotterdam, Department of Neuroscience, Dr. Molewaterplein 50, 3015 Dr Rotterdam, The Netherlands

Stanislas Dehaene INSERM Unité de Recherche 562, Service Hospitalier Frédéric Joliot; CEA, 4, Place du Général Leclerc, F-91401 Orsay, France

Dori Derdikman Department of Neurobiology, The Weizmann Institute of Science, Rehovot, 76100, Israel

Mathew E. Diamond Cognitive Neuroscience Sector, International School for Advanced Studies, SISSA, Via Beirut 2-4, 34014 Trieste, Italy

Jochen Ditterich Center for Neuroscience, 1544 Newton Court, University of California, Davis, CA 95616, USA

Hossein Esteky Shaheed Behheshti School of Medicine and School of Cognitive Sciences (SCS), Institute for Studies in Theoretical Physics and Mathematics (IPM), Niavaran, PO Box 19395-5746, Tehran, Iran

Joshua I. Gold Department of Neuroscience, , University of Pennsylvania, 116 Johnson Pavilion, Philadelphia, PA 19104-6074, USA

Patrick Haggard Institute of Cognitive Neuroscience, University College London, 17 Queen Square, London WC1N 3AR, UK

Justin A. Harris School of Psychology, University of Sydney, Sydney, New South Wales 2006, Australia

Uri Hasson   Room 955, Meyer Bldg, Center for Neural Science, New York University, 4 Washington Place, New York, NY 10003, USA

Leah Krubitzer   Center for Neuroscience, Department of Psychology, 1544 Newton Court, Davis, CA 95616, USA

Nikos K. Logothetis   Max Planck Institute for Biological Cybernetics, Rm L103, Spemannstraße 38, 72076 Tübingen, Germany

Jacques Mehler   Cognitive Neuroscience Sector, International School for Advanced Studies, SISSA, Via Beirut 2-4, 34014 Trieste, Italy

Carlo A. Porro   Dip. Scienze Biomediche, Sezione Fisiologia, Univ. di Modena e Reggio Emilia, Via Campi 287, I-41100 Modena, Italy

Giacomo Rizzolatti   Dipartimento di Neuroscienze, Sezione di Fisiologia, Università di Parma, Via Volturno, 39E, Parma, I-43100, Italy

Ranulfo Romo *(Chair)*   Instituto de Fisiologia Celular, Universidad Nacional Autonoma de Mexico, Apartado Postal 70-253, 04510 Mexico Distrito Federal, Mexico

Raffaella I. Rumiati   Cognitive Neuroscience Sector, Scuola Internazionale Superiore di Studi Avanzati, Via Beirut 2-4, Trieste, 34014, Italy

Jeffrey D. Schall   Center for Integrative and Cognitive Neuroscience, Vanderbilt Vision Research Center, Department of Psychology, Vanderbilt University, Nashville, TN 37203, USA

Stephen H. Scott   Department of Anatomy & Cell Biology, CIHR Group in Sensory-Motor Systems, Centre for Neuroscience Studies, Queen's University, Kingston, Ontario, K7L 3N6, Canada

David L. Sparks   3400 Connell Drive, Pensacola, Florida 32503, USA

Manabu Tanifuji   Laboratory for Integrative Neural Systems, RIKEN Brain Science Institute, 2-1 Hirosawa, Wako, Saitama, 351-0198, Japan

Alessandro Treves   Programme in Neuroscience, SISSA, Via Beirut 2-4, 35014 Trieste, Italy

Manos Tsakiris *(Novartis Foundation Bursar)* Institute of Cognitive Neuroscience, University College London, 17 Queen Square, London, WC1N 3AR, UK

Daniel M. Wolpert Sobell Department of Motor Neuroscience and Movement Disorders, Institute of Neurology, University College London, Queen Square, London WC1N 3BG, UK

Chair's introduction

Ranulfo Romo

Instituto de Fisiología Celular, Universidad Nacional Autónoma de México, 04510 México, DF, México

I am pleased to be attending this meeting and to participate in this opening programme. Firstly, because I will have the opportunity to hear about the latest results and thoughts on the neural mechanisms of percept, decision-making and action. Secondly, because I am sure there will be discussions about the progress made not only on each of these cognitive functions, but importantly on what neural operations link them. In other words, we will be working towards an integrated understanding of the neural processes that link sensation to action.

Over the past several years, knowledge has accumulated dramatically, in such a way that it is almost impossible to track new experimental findings and thoughts on each of these themes. This symposium is based on a proposal made by Professor Mathew Diamond, who certainly perceived the necessity for a synthesis of these themes and I thank him for his initiative and effort. Let me summarize what I believe are the main themes that will be discussed in this three day meeting.

A central issue in neuroscience is the elucidation of how sensory information is used to generate behavioural actions. In principle, this process can be understood as a chain of three basic neural operations. The representation of the physical attributes of the environment (sensory coding) and the execution of motor commands (motor coding) can be regarded as the end points of this chain of operations. In the middle of the chain there is a crucial processing step in which the sensory representations are analysed and transformed in such a way that the nervous system is able to choose the adequate motor response from an enormous repertoire of possible behavioural responses.

A heuristic change in the study of percept, decision-making and action has been the realization that the study of sensory performance called psychophysics, and the neural events triggered by sensory stimuli, called sensory neurophysiology, are simply different experimental approaches to the same set of problems. Most of the participants in this symposium have carried out investigations in which they measured both sensory performance and neural events, and sought explanations of the former in terms of the latter. Thus the neurophysiologist aims to characterize brain events in a behavioural context allowing measurement of the sensations in terms of detection, discriminations, categorizations and motor actions.

1

The underlying belief behind this approach is that unravelling the neural representations of sensory stimuli from the periphery to early stages of cortical processing is key to addressing brain function, be it local or distributed. Investigations in several systems have shown how neural activity represents the physical parameters of sensory stimuli in both the periphery and central areas of the brain. These results have paved the way for new questions that are more closely related to cognitive processing. For example, how are the neural representations of sensory stimuli related to perception? What attributes of the observed neural responses are relevant for downstream networks and how do these responses influence decision-making and behaviour? To understand the neural dynamics of decision-making we first need to know how the physical variables on which a decision is based are encoded. One of the main challenges of this approach is that even the simplest cognitive tasks engage a number of cortical areas, and each one could render sensory information in a different way, or combine it with other types of stored signals representing, for example, past experience or future actions.

As will be seen during this symposium, studies in behaving animals that combine psychophysical and neurophysiological experiments have provided new insights into this problem. In particular, there has been important progress regarding how neural codes are related to perception and decision making in the somatic and visual systems. The basic philosophy of this approach has been to investigate cognitive tasks using highly simplified stimuli, so that diverse subcortical and cortical areas can be examined during the same behaviour. The idea is that if neural codes for such stimuli are readily identifiable, then determining the individual functional roles of those areas should become less difficult.

Anatomical studies have shown that sensory cortices are connected with motor areas of the frontal lobe and subcortical structures as well. The question that arises then is whether there is a truly clear distinction between those areas presumably dedicated to sensory processing and those traditionally viewed as motor. There are two possibilities. First, the motor areas could process a fully formed decision signal in order to generate an appropriate set of motor commands. In this case, information and processes used before reaching a decision should be mostly absent from motor activity. Second, the motor areas could participate more actively in the decision process, in which case, they should reflect the sensory inputs regardless of the motor outputs.

Although our understanding is still far from complete, a sufficient foundation of knowledge now exists about different levels of sensory and motor systems to support direct investigation of higher representations, which lack robust sensory or motor qualities and instead are closely related to cognitive processes. Current research in this area asks questions regarding the nature of specialized neuronal signals underlying perceptual decisions, motor plans, emotions, the representation of reward and language. Some of the papers that will be presented at this symposium are part of an effort to answer these questions.

To close, I would like to extend my sincere thanks to Derek Chadwick of the Novartis Foundation for his admirable work. Dr Chadwick and his assistant Jane Dempster did all that is necessary for the flow and exchange of ideas and results at this meeting. Professor Mathew Diamond, our host, carried on his shoulders all the logistic of this meeting. To him and to the Novartis Foundation we owe our thanks for this opportunity to learn and work together at this symposium.

Active construction of percepts about object location

Dori Derdikman, Marcin Szwed, Knarik Bagdasarian, Per Magne Knutsen, Maciej Pietr, Chunxiu Yu, Amos Arieli and Ehud Ahissar

Neurobiology Department, Weizmann Institute of Science, Rehovot, Israel

Abstract. Mammals acquire much of their sensory information by actively moving their sensory organs. Rats, in particular, scan their surrounding environment with their whiskers. This form of active sensing induces specific patterns of temporal encoding of sensory information, which are based on a conversion of space into time via sensor movement. We investigate the ways in which object location is encoded by the whiskers and decoded by the brain. We recorded from first-order neurons located in the trigeminal ganglion (TG) of anaesthetized rats during epochs of artificial whisking induced by electrical stimulation of the facial motor nerve. We found that TG neurons encode the three positional co-ordinates with different codes. The horizontal coordinate (along the backward–forward axis) is encoded by two encoding schemes, both relying on the firing times of one type of TG neuron, the 'contact cell'. The radial coordinate (from face outward) is encoded primarily by the firing magnitude of another type of TG neurons, the 'pressure cell'. The vertical coordinate (from ground up) is encoded by the identity of activated neurons. The decoding schemes of at least some of these sensory cues, our data suggest, are also active: cortical representations are generated by a thalamic comparison of cortical expectations with incoming sensory data.

2005 Percept, decision, action: bridging the gaps. Wiley, Chichester (Novartis Foundation Symposium 270) p 4–17

We can define active sensing as a sensory strategy that is based on sensory acquisition via moving senses (Gibson 1962, Ahissar & Arieli 2001). In both vision and touch, sensation is mediated by moving arrays of change-sensitive receptors across external objects (Fig. 1). In both cases movement of the sensory apparatus is an integral part of sensation. The rat possesses such an active sensing system too— the whisker somatosensory system (Fig. 2) composed of an array of large whiskers, or macrovibrissae and well-defined neural structures. While scanning its environment, the rat moves its whiskers back and forth in a rhythmic fashion at about 5–10 Hz. The question which we are trying to understand is how the movement of the whiskers affects sensation.

Take for example a rat, while it is trying to discriminate the position of two objects, in a two-forced choice discrimination task. The rat, engaged in the task, has

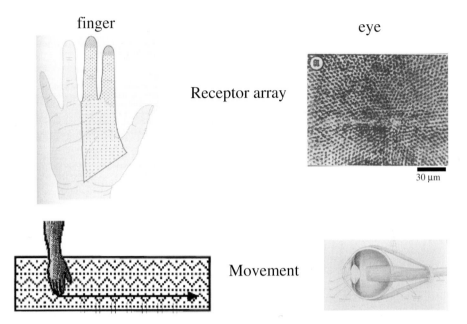

FIG. 1.   Active sensing systems in humans. Receptor arrays in the finger, or in the retina, are complemented by active movements of the hand and of the eye. Top right panel Young (1971), bottom left modified from Ahissar (1998).

to discriminate which of the two poles is nearer to its home box (Knutsen et al 2003, 2005a). The rat moves its whiskers and head in order to solve the task. We use the term motor-sensory here, rather than the traditional term sensory-motor, in order to emphasize the fact that motor movements induce sensations (Fig. 3).

The rat trigeminal and motor system is one example of such a motorsensory loop (Kleinfeld et al 1999, Ahissar & Kleinfeld 2003) (Fig. 4). In the outer loop, information flows from the vibrissae, via the trigeminal ganglion (TG) first-order neurons, to the brainstem and from there to the thalamus and up to the cortex. Motor information flows from the cortex to the brainstem motor nuclei back to the vibrissae, thus closing the loop. Several parallel pathways also close the loop at lower levels. Active sensing in this system is concerned with the discussion of how the movements of the vibrissa affect sensation, and how the movements are used for sensation.

We focus on two stations in the motorsensory transformation. First, we describe the input signals of the first-order neurons in the TG, representing the first station of sensory encoding (Szwed et al 2003). Next, we will focus on the vibrissa itself, and will try to decipher whisker kinematics, in order to understand the types of movements that generated the sensory encoding seen in the first-order sensory neurons in the trigeminal ganglion.

Receptor
array

Movement

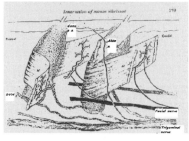

FIG. 2. Active sensing system in the rat. Receptor array is whisker pad. The rat moves its whiskers actively to scan the environment, by a set of specialized ('intrinsic') muscles. Upper two figures: www.neurobio.pitt.edu/barrels. Reproduced by permission of Blackwell Publishing. Bottom figure from Dorfl (1982).

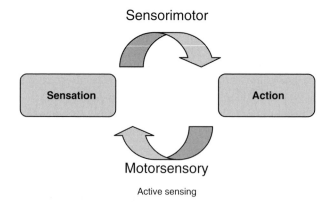

FIG. 3. Sensorimotor transformation between sensation and action is complemented by motorsensory transformation between action and sensation.

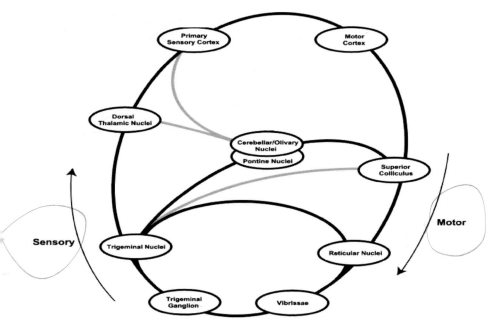

FIG. 4. The rat trigeminal system sensory-motor-sensory loop. Reproduced courtesy of David Kleinfeld, from Kleinfeld et al (1999) with permission from Taylor & Francis Ltd (www.tandf.co.uk/journals).

The structure of the input signals in the trigeminal ganglion

Szwed et al (2003), have modified and re-introduced artificial whisking as a method for investigating active-sensing (the method was first applied in the 1960s [Zucker & Welker 1969]). In artificial whisking, the facial motor nerve innervating the whiskers was electrically stimulated to generate stereotypical whisker protraction and retraction movements, at a rate of 5 Hz, which is in the physiological range of natural whisking. To determine the structure of the input signals Szwed et al (2003) applied artificial whisking in the anaesthetized animal, and recorded from the trigeminal ganglion while the anaesthetized rat was artificially whisking in free air or against an object (vertical steel pole) located at different positions. Note that the motor-sensory-motor loop was opened in this paradigm by cutting the connection between the motor nerve and the motor nuclei[1] (Fig. 5). During active touch, trigeminal ganglion neurons presented a rich repertoire of responses, which could not be

---

[1] This is in contradistinction with passive mechanical whisker stimulation, where the motor–sensory–motor loop is opened at a different point—at the interface between the muscles and the whiskers.

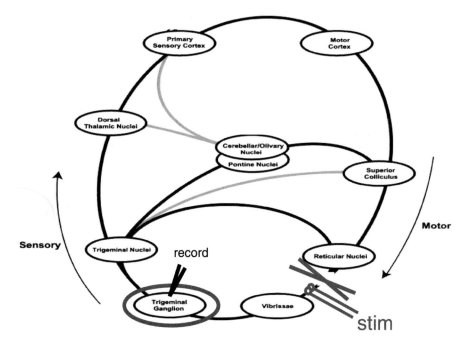

FIG. 5. Description of the paradigm used in Szwed et al (2003), in relation to the sensory–motor–sensory loop. Reproduced from Kleinfeld (1999) with permission from Taylor & Francis Ltd.

predicted from their responses to passive stimuli. When recording extracellularly in the TG, they found three major classes of neuronal responses: first, there were cells that responded to the whisking of the rat (Fig. 6). Those cells did not fire when the whiskers were stationary, and locked their response to the rat's whisking during electrical stimulation. The cells were indifferent to obstacles presented in the path of the whisker—they responded in a similar manner when the rat was whisking in free air, and when it was whisking while a vertical object was presented in the path of the whiskers. They termed these cells *whisking cells*. When looking at single examples of whisking cells, one can see that various neurons responded at various phases of the whisking. One responded at the onset of protraction, another at a later phase of the protraction cycle; a third neuron responded phasically both during protraction and during retraction, a fourth neuron responded tonically during the entire protraction phase (Fig. 7).

The second group of cells observed was *touch cells*. These cells did not respond when the rat was whisking. However, they responded when an obstacle was introduced in the path of the whiskers, near the distal end of the whiskers. Three subclasses of touch neurons were found: *contact* neurons, which responded phasically only during the moment of contact, *pressure* neurons, which responded tonically

Whisking cells
respond only to whisking

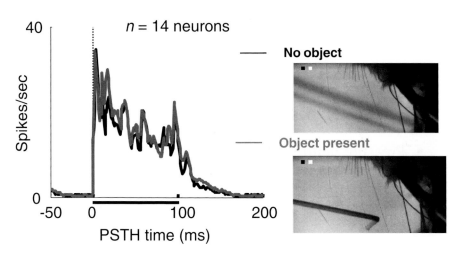

FIG. 6. Response of trigeminal-ganglion whisking cells in object and no-object conditions. Reprinted from Szwed et al (2003) with permission from Elsevier.

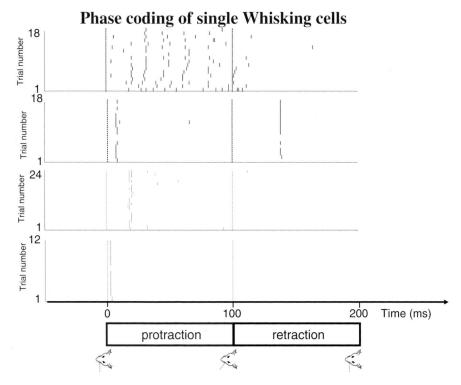

FIG. 7. Phase coding of whisking cells. Four examples of trigeminal ganglion whisking cells responding at various phases of the whisking cycle. Reprinted from Szwed et al (2003) with permission from Elsevier.

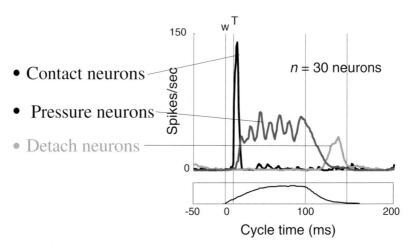

FIG. 8.   Different classes of touch cells. Reprinted from Szwed et al (2003) with permission from Elsevier.

while the whisker was pressing against the object, and *detach* neurons, which responded phasically only when the whisker was detaching from the object in its path (Fig. 8).

There also existed a third class of cells, in between the other two classes: The *whisking-touch* neurons, which responded to whisking without object, however, demonstrated a more vigorous response when an object was present in the path of the whiskers (Fig. 9).

Whisker kinematics and their effect on encoding

In order to understand what is encoded during active sensing, one must investigate the physical kinematics of the sensory receptor movements, and the interactions of these receptors with the external world. In the case of the rat whisker system, this starts with the kinematics of the external shaft of the whisker, and is followed by transformations from the external to internal shaft, and by mechanoreceptive transduction. We started with the kinematics of the external shaft, and asked whether it could explain the selectivity in responses of whisking and touch neurons.

Since all first-order neurons respond to passive stimuli, two questions arise immediately: First, why don't all receptors respond to whisking in air? That is, why do touch neurons respond to an object, but do not respond to whisking without an object? Second, why don't all receptors respond to touch? In other words, why do

Whisking/Touch cells
respond to whisking and touch

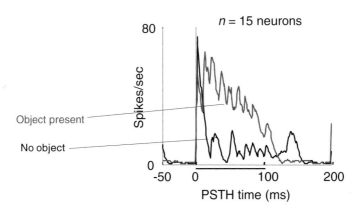

FIG. 9.   Response of whisking–touch cells. Reprinted from Szwed et al (2003) with permission from Elsevier.

whisking cells not increase their response when an object is presented in the path of the whiskers? Do we have some indications as to why TG neurons show selectivity to different classes of stimuli, as described above? A partial answer is given by looking at the structure of the whisker follicle (Vincent 1913, Rice et al 1986, Ebara et al 2002) (Fig. 10). The follicle is an extremely elaborate structure, containing hundreds of receptors of different kinds, which are organized at different positions inside the follicle. It contains large blood sinuses, which probably help to create a mechanically isolated environment within it (Mitchinson et al 2004). The follicle is surrounded by an intrinsic muscle that can move it when the rat is whisking (Dorfl 1982). We hypothesize that different receptors, located at different positions inside the follicle, are selective to different types of stimuli. For example, receptors residing in the central part, which is mechanically isolated from the environment by the blood sinus, are probably related to touch neurons, described in the previous section, because they are perhaps not affected by the movement of the muscles while the rat is whisking, however they are affected by vibrations of the whisker shaft. On the other hand, receptors residing near the mouth of the follicle are not isolated mechanically in the same way, and thus sense the movements of the follicle. We expect that receptors near the follicle mouth will be more of the whisking type.

A second clue towards answering the question of the selectivity of first-order neurons can come from the analysis of whisker kinematics (Derdikman et al 2003). We analysed whisker kinematics in detail by tracking the whiskers of an artificially whisking rat using a semi-automatic whisker tracking algorithm (Knutsen et al

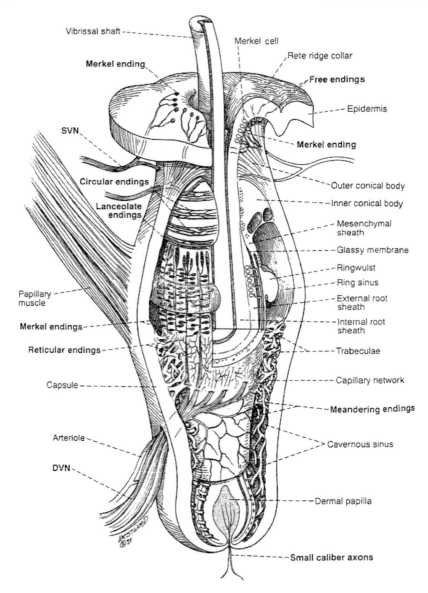

FIG. 10.   Illustration of the whisker follicle of the rat. Reproduced by permission of John Wiley and Sons Inc, from Rice et al (1993).

2005b). We measured the angle of various whiskers as a function of time, while the whiskers of the rat were obstructed by an object positioned in their path. Preliminary results show that the selectivity of whisking and touch neurons may be explained on the basis of whisker kinematics (Derdikman et al 2003).

Summary

First we investigated the responses of neurons in the TG, and demonstrated (Szwed et al 2003) that there are various classes of neurons, selective to various components of the response. There are whisking neurons, which respond to free air whisking but respond to obstacles, touch neurons, which care only about obstacles in the path of the whiskers, but do not care about the whisking itself, and the intermediate group, the whisking–touch neurons, which signal about both whisking and touch. Next, we investigated the kinematics of whiskers whose physical movements are known to affect the encoding of the first-order neurons. Preliminary results suggest (Derdikman et al 2003) that whisker kinematics can explain the selectivity of whisking and touch neurons.

References

Ahissar E 1998 Temporal-code to rate-code conversion by neuronal phase-locked loops. Neural Comput 10:597–650
Ahissar E, Arieli A 2001 Figuring space by time. Neuron 32:185–201
Ahissar E, Kleinfeld D 2003 Closed-loop neuronal computations: focus on vibrissa somatosensation in rat. Cereb Cortex 13:53–62
Derdikman D, Knutsen PM, Ahissar E 2003 Integrating kinematic data and extracellular recordings in the 'electrically whisking' rat: effects of whisker angle and curvature. Barrels XVI Meeting, New Orleans, LA
Dorfl J 1982 The musculature of the mystacial vibrissae of the white mouse. J Anat 135:147–154
Ebara S, Kumamoto K, Matsuura T, Mazurkiewicz JE, Rice FL 2002 Similarities and differences in the innervation of mystacial vibrissal follicle-sinus complexes in the rat and cat: a confocal microscopic study. J Comp Neurol 449:103–119
Gibson JJ 1962 Observations on active touch. Psychol Rev 69:477–491
Kleinfeld D, Berg RW, O'Connor SM 1999 Anatomical loops and their electrical dynamics in relation to whisking by rat. Somatosens Mot Res 16:69–88
Knutsen PM, Pietr M, Derdikman D, Ahissar E 2003 Whisking behavior of freely-moving rats in an object localization task. Neural Plasticity 11:1
Knutsen PM, Pietr M, Ahissar E 2005a Trajectory control during rodent whisking. Neural Plasticity 12:32
Knutsen PM, Derdikman D, Ahissar E 2005b Tracking whisker and head movements in unrestrained behaving rodents. J Neurophysiol 93:2294–2301
Mitchinson B, Gurney KN, Redgrave P et al 2004 Empirically inspired simulated electromechanical model of the rat mystacial follicle-sinus complex. Proc Roy Soc Lond Ser B Biol Sci 271:2509–2516
Rice FL, Mance A, Munger BL 1986 A comparative light microscopic analysis of the sensory innervation of the mystacial pad. I. Innervation of vibrissal follicle-sinus complexes. J Comp Neurol 252:154–174

Rice FL, Kinnman E, Aldskogius H, Johansson O, Arvidsson J 1993 The innvervation of the mystacial pad of the rat as revealed by PGP 9.5 immunofluorescence J Comp Neurol 337:366–385

Szwed M, Bagdasarian K, Ahissar E 2003 Encoding of vibrissal active touch. Neuron 40:621–630

Vincent SB 1913 The tactile hair of the white rat. J Comp Neurol 23:1–36

Young RW 1971 The renewal of rod and cone outer segments in the rhesus monkey. J Cell Biol 49:303–318

Zucker E, Welker WI 1969 Coding of somatic sensory input by vibrissae neurons in the rat's trigeminal ganglion. Brain Res 12:138–156

DISCUSSION

Diamond: In your data you showed that the sensory system acting by itself carries a lot of information about the position of the object that the rat has touched with its whiskers. Yet none of that information actually requires the motor signal itself. Would you guess that the rat somehow uses the motor output as well, and combines this with sensory information to localize objects?

Derdikman: If you mean some kind of corollary discharge in this system, I wouldn't know how to answer that at this stage. I think the intrinsic muscles do not contain proprioceptors. At least some proprioceptive information is lacking in this system.

Diamond: There is a central pattern generator in the brainstem that produces the whisking cycle, but people haven't looked much about how this feeds back into the sensory system.

Treves: Did you mean to say that there is a bimodal clustering in the ganglion, between units that respond to touch and units that respond only to whisking, or else that there is a continuum, of which you showed the extremes?

Derdikman: The basic idea is that there is a continuum with very strong extremes. The populations which are related to the extremes are larger than the populations that are related to the continuum.

Logothetis: At one point you posed the question of why there are both whisker-sensing and touch-sensitive cells in the trigeminal pathway. It is not clear to me that you answered this question. It seems to me that if you are sensing curvature then you don't need the touch cells at all. There are three kinds of movement I could see as a naïve observer. One is where the angle of velocity of every point in the whisker is constant. Then there are those where there is an oscillatory pattern, and thirdly those which clearly change curvature, where every point on the whisker does not have the same angle of velocity. This third category should be able to notify the system that there is an object there. Is this correct?

Derdikman: This is one category that may notify the system of an object. Of course, what is lacking in this work is the actual demonstration of correlation between the various kinematic parameters and the responses in the ganglion. This is work which is in progress. I would predict that the curvature is related to the subpopulation of pressure cells. The curvature remains high all along the contact between the whisker and the object, just like the pressure cells have a more tonic response. Many of the cells in the trigeminal ganglion, however, are phasic. To find the correct kinematic parameter for such cells, we should look at a kinematic parameter such as square velocity, related to the dissipation of energy during the collision between the whisker and the object. I was demonstrating that it is possible to find such kinematic parameters which can assist us in showing the selectivity between different types of responses seen inside the follicle, but I do not commit to a specific parameter more than to others at this stage.

Logothetis: In what you showed in the plots, where you have angle, the angle is just a base angle. This means those base angles for the whiskers that are touching objects should decrease.

Derdikman: Contrary to intuition, when an object is touching the whiskers at a distal position, the base angle is hardly affected, and the change in curvature dominates. The work I showed was related only to horizontal object localization, when the object is touching the whisker at a distal location. Things are different when an object is touching at a proximal position. In such cases indeed the angle is smaller when the object is touching the whisker.

Sparks: Were all the data that we saw obtained when whisking was induced by electrical stimulation?

Derdikman: Yes.

Sparks: Have you compared responses and categories during active whisking versus stimulation-evoked whisking?

Derdikman: Yes. One fundamental difference is that when we look at passive as opposed to active stimulation, usually all the neurons respond to the stimulation. The selectivity emerges mostly when we do an active-whisking paradigm. The passive-stimulation paradigm doesn't usually create such a selectivity, as seen in the case of active-whisking, where there is such selective responses between whisking and touch.

Albright: Does the whisking frequency vary in natural conditions? And over what kind of range?

Derdikman: Yes, it varies at a range of about 5–20 Hz. Our paradigm is 5 Hz, which is at the lower range of the whisking frequencies.

Albright: Is there any evidence that the differences in frequency might affect the dynamics of the responses?

Derdikman: Perhaps, however the basic separation into touch cells and whisking cells remains also at higher frequencies (Szwed 2003).

Brecht: The animal keeps the frequency remarkably constant in any burst of whisking. When they do another burst it might be different. This makes me think that there is some significance to the rhythmicity.

Albright: There might be different frequencies for different purposes.

Gold: In additional to controlling the whisking frequency under natural conditions, how important are movements of the head in terms of this active process that might influence neurons that are involved in sensory–motor integration? The exploratory part would have to be involved at least as much as changing of frequency of whisking. Then, under those more natural conditions, how might you expect those kinds of signals to affect the properties of these neurons?

Derdikman: To answer this question, we have to record from trigeminal ganglion neurons in an awake-behaving rat. My colleagues, Per Knutsen and Maciej Pietr, have looked at hundreds of movies of rats while they are whisking. They have tracked their whiskers, using an automated whisker-tracking program (Knutsen et al 2005a). In the task I described of discrimination of the position of the object, the rat is usually slowly approaching it. As it does this, it slowly reduces the amplitude of the whisking. Usually there is an amplitude–frequency relationship, so as the rat reduces the amplitude it increases the frequency of whisking (Knutsen 2005b).

Haggard: Do we know how those movements then change after the whisker has touched an object? This gets back to the idea of exploration: do we know how this happens naturally? I am struck in your work by how you have solved a problem frequently encountered in human work on active touch, which is that we can't control how humans explore objects. It is therefore difficult to do controlled experiments on human active touch. By providing electrical stimulation-evoked movements you could solve this problem. The question would be, after touching an object how does a rat then proceed to explore the object with its whiskers under natural circumstances.

Derdikman: Again, following observations by Knutsen and Pietr, as the rat approaches an object, after it touches the first time it will usually touch several more times. There may be some motor feedback, regulating the size of the whisks. The first whisk will be a reaching movement, and proceeding this there will be several additional events of object touch (Pietr et al 2004).

Diamond: At a recent meeting I saw data from your laboratory that deal with this question (Pietr et al 2004). When the rat touches the vertical pole with one whisker, this whisker is blocked but the others seem to move forward more quickly. Their impression was that this might indicate a fast feedback loop where resistance against the whisker induces additional movement.

Derdikman: That is one possibility. A second possibility is that the effects seen are mostly mechanical and not related to active sensory feedback. We're currently investigating this question. A related question is how does this curvature parameter change in the awake-behaving rat? Does the rat continue to push the object, increasing the curvature, or does it save some energy by stopping the movement of the whiskers once they are touching the object? Perhaps there is some feedback, such that the rat will stop the whisker movement when the whisker is touching the object, while continuing to move the other whiskers.

References

Knutsen PM, Derdikman D, Ahissar E 2005a Tracking whisker and head movements in unrestrained behaving rodents. J Neurophysiol 93:2294–2301

Knutsen PM, Pietr M, Ahissar E 2005b Trajectory control during rodent whisking. Neural Plasticity 12:32

Pietr M, Knutsen PM, Derdikman D, Ahissar E 2004 Whisking kinematics in unrestrained rats and mice: signs of closed loop control. Gordon Research Conference on 'Sensory coding and the natural environment', Oxford, UK

Szwed M, Bagdasarian K, Ahissar E 2003 Encoding of vibrissal active touch. Neuron 40: 621–630

Neuronal encoding of natural stimuli: the rat tactile system

Mathew E. Diamond, Erik Zorzin and Ehsan Arabzadeh

Cognitive Neuroscience Sector, International School for Advanced Studies, Via Beirut 2/4 34014 Trieste, Italy

Abstract. A major challenge of sensory systems neuroscience is to quantify the brain activity underlying perceptual experiences and to explain this activity as the outcome of elemental neuronal response properties. One strategy is to measure variations in neuronal response in relation to controlled variations in an artificial stimulus. The limitation is that the stimuli scarcely resemble those which the sensory system has evolved to process—natural, behaviourally relevant stimuli. A more recent strategy is to measure neuronal responses during presentation of natural stimuli, but such experiments have failed to predict the observed responses according to the fundamental properties of neurons. In the work described here, we focus on tactile sensation in rats, and try to bridge the gap between neurons' responses to natural stimuli and their responses to controlled, artificial stimuli. We focus on texture, a submodality in which the rat whisker sensory system excels. Because the physical characteristics of texture stimuli have not yet been studied, the first set of experiments measures textures from the whiskers' point of view. The second set of experiments describes neurons' responses to textures. The third set of experiments computes kernels (estimates of the extracted stimulus features) of sensory neurons using white noise and then tries to account for natural texture responses according to these kernels. These investigations suggest ways of using natural stimuli to assemble a more complete picture of the neuronal basis of tactile sensation.

2005 Percept, decision, action: bridging the gaps. Wiley, Chichester (Novartis Foundation Symposium 270) p 18–37

Linking neuronal activity to sensation

This chapter concerns the problem of how the brain builds up an internal representation of the external world. Such internal representations constitute the neuronal basis of perceptions and are the substance from which judgements and decisions are made. The specific problem addressed here is the neuronal basis of the sense of touch.

What is the neuronal language of sensation? How can neuroscientists discover which features of a neuronal population's activity produce the experience of one sensation and not another? Measurements of neuronal responses to simplified stimuli—e.g. sinusoidal light gratings or pure frequency tone pips—can provide a complete description of neuronal feature extraction properties (Carandini et al 1997, Adelman et al 2003), but they do not reveal the brain activity underlying normal perceptual experiences. Another approach is to measure neuronal activity during natural stimuli (i.e. visual scenes or animal calls). Here, the drawback is that the features evoking spikes during natural stimulation are complex and difficult to quantify (Theunissen et al 2000, Simoncelli & Olshausen 2001). Although responses can be recorded, the investigator cannot know exactly what neuronal processing operation led to the observed firing sequence.

In principal, one can bridge the gap between artificial and natural stimuli by:

- measuring neuronal activity during ecologically relevant stimuli
- using artificial stimulation to construct 'kernels' (estimates of the stimulus features extracted by neurons), and
- applying the kernels to the natural stimuli to test whether they predict the observed response.

In the experiments described here, we have tried to form such a bridge in order to formulate a more complete picture of tactile sensation in rats. We focus on texture since the sensory system has extraordinary capacities in this submodality. Other work presented in this volume (Derdikman et al 2005, this volume), concerns the neuronal representation of *object location* in the whisker sensory system. These two problems—*where* and *what* is the object (what are its features, e.g. its texture)—are perhaps the somatosensory analogue to the *where* and *what* problems that the visual system must solve, also discussed in this volume.

Because the physical parameters of texture stimuli have yet to be studied, the first set of experiments measures how whiskers and textures interact. The second set of experiments describes neurons' responses to texture stimuli. The third set of experiments computes the kernels of sensory neurons and then tries to account for natural texture responses according to them. Simulations show that during the texture encoding, neurons are in fact operating on natural stimuli according to these kernels.

The whisker system

Rats are nocturnal animals and depend on their vibrissal sensory system for navigation, object localization, judgement of the roughness or texture of surfaces, and the size and shape of small objects. In particular, humans and rats have roughly equivalent capacities in texture discrimination (Carvell & Simons 1990). The whisker sensory system is efficient and highly evolved; as such, we believe that insights into its functioning will generalize to other specialized systems, like the primate visual system.

The whisker is a specialized hair, and the transduction process occurs within the whisker follicle (Fig. 1A). The neural pathway begins at the receptors, the primary afferent terminations resting on the whisker shaft (Ebara et al 2002). Whisker deflections or vibrations cause a stretching of the membrane of the primary afferent fibre, depolarizing the terminal and inducing a train of action potentials that stream along the afferent nerve. Each follicle contains about 200–300 sensory fibres.

Sensory signals travel along the afferent nerve, past the trigeminal ganglion, to the trigeminal nuclei in the brain stem (Fig. 1B). Here the first synapse is located. The axons of second-order neurons cross the brain midline in the medial lemniscus and travel to the thalamic somatosensory nuclei, where the second synapse is located. Thalamic neurons project to the primary somatosensory cortex, conveying information mainly to layer IV targets (Diamond 1995, Ahissar et al 2000).

The whisker area of somatosensory cortex is one of the clearest examples of mammalian columnar organization. Woolsey & Van der Loos (1970) discovered that in mouse discrete clusters of small cells, 'barrels', could be delineated in a tangential section through layer IV, and that the spatial arrangement of clusters replicated the spatial arrangement of the large facial whiskers on the opposite side of the snout. Each whisker has a name, e.g. C1, and a corresponding cortical 'barrel', C1. A similar organization exists in rats (Welker 1971).

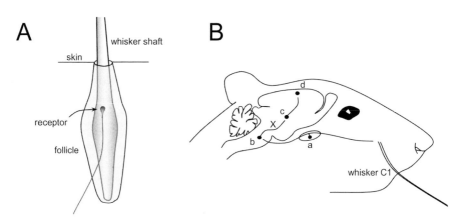

FIG. 1.   Essential features of the whisker sensory system. (A) Sensory transduction is accomplished by a mechanoreceptor, a nerve termination resting on the whisker shaft within the whisker follicle. (B) Simplified scheme of the sensory pathway originating in one whisker, C1, on the right side of the rat's snout. Impulses from the receptor are transmitted past the cell body in the right trigeminal ganglion (a). The second neuron is located in the trigeminal nucleus of the brain stem (b) and its axon crosses the midline of the brain (X) to the left thalamus (c). The thalamic neuron projects to left 'barrel cortex' (d).

What stimulus is 'natural' for this system?

Since the remarkable discovery of the whisker-to-barrel pathway 34 years ago, the circuitry of the sensory pathway has been studied in the smallest detail. We know with good precision, for example, the synaptic connectivity within and between cortical columns (Thomson & Bannister 2003). Yet, we know almost nothing about neuronal activity during natural sensory events. To begin with, what is a *natural* stimulus? In visual system we can describe the stimulus either according to its physical properties (photons and wavelengths; Baylor et al 1984) or its more holistic properties (natural scenes and their statistics; Kayser et al 2003). However, we are not yet able to formulate a quantitative description of this sort for the whisker system because object exploration in the tactile modality derives from active palpation (in this respect, tactile sensation in rats resembles that in humans). Rather than passively allowing objects to encounter the whisker, as the visual system accepts images that fall upon the retina, rats sweep their whiskers across surfaces in a rhythmic forward–backward cycle. And it is impossible to know, *a priori*, the object/whisker interaction and, therefore, the signal that arrives in the whisker follicle. Our first task was to achieve a first-order understanding of whisker signals.

There have been no reports concerning the neuronal activity generated by whisking along irregular surfaces, and the differences in activity associated with two surfaces remain unknown. As our second task, we tried to produce a preliminary description of neural coding of textures. By measuring the activity of ganglion neurons, we estimated how the sweeping motion of whiskers along a surface is converted to a neuronal impulse code. By measuring activity in the cortex, we explored the neuronal representation that rats must rely on to judge the identity of external objects (Whitfield 1979, Hutson & Masterton 1986, Guic-Robles et al 1992).

Kinetic signatures induced by textures

The experiments began with the collection records of the natural movement of whiskers across surfaces. We electrically stimulated the facial nerve to generate 8 Hz whisking movements that closely resemble whisker trajectories in awake rats (Brown & Waite 1974, Bermejo et al 2002, Szwed et al 2003). Meanwhile, whisker displacements transmitted to the receptors in the follicle were measured by an optical sensor placed 1 mm from the skin (Fig. 2A,B). The vertical and horizontal channels of the sensor reported whisker position with $<3\,\mu$m spatial and 0.13 ms temporal resolution. Later, these became the stimulus set to probe the neuronal representation of texture (Fig. 2C,D).

We measured movements under different conditions: whisking with no object contact ('free whisking'), and whisking on a number of different surfaces. The surface was oriented so that the whisker remained in contact during the entire whisk

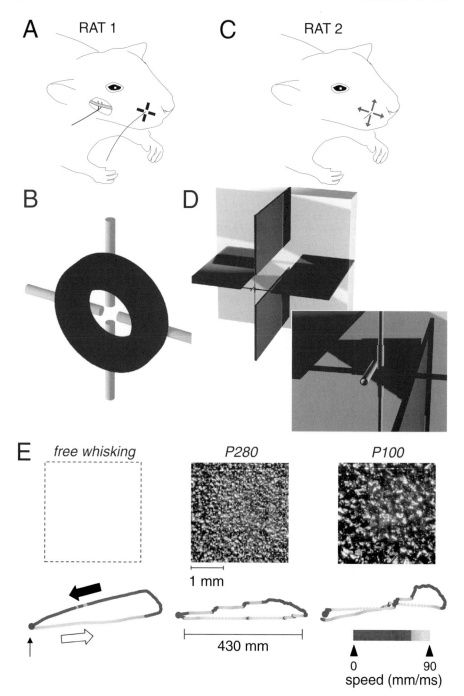

A RAT 1

C RAT 2

B

D

E *free whisking* P280 P100

1 mm

430 mm

0   90
speed (mm/ms)

trajectory. From this library, we illustrate free whisking as well as two sandpapers of different coarseness, denominated P280 and P100 (Fig. 2E, upper panel). Figure 2E (lower panel) shows the trajectory of the base of the whisker, in position (given by x–y coordinates) and speed (given by greyscale), under these three conditions. During free whisking (left side), the trajectory was a smooth ellipsoid, the principal axis aligned with protraction and retraction movements. In contrast, whisking across grainy surfaces produced irregularities in the trajectory—jumps, stops and starts, and bursts of high and low velocity (Fig. 2E, middle and right). Most importantly, the whisker shaft movements associated with the two textures were distinct from each other. Each whisking condition was associated with a unique kinetic signature.

Neuronal encoding of kinetic signatures

The critical issues, then, are:

- to quantify whisker vibrations; and
- to determine whether they induce neuronal representations that could serve as the basis for fine sensory discriminations

First, in Figure 3A the 'kinetic signatures' are presented again, this time in the form of velocity profiles—the temporal sequence of velocity features across the course of a whisk. Each velocity profile covers 250 ms (two complete forward/backward whisks) and consists of two histograms—horizontal (V_H) and vertical (V_V) velocity. From these profiles, we conclude that the velocity features that make each texture unique include the number, magnitude, duration, and temporal spacing of the peaks in V_H. V_V varied less across textures.

◀

FIG. 2. Collection of texture library and stimulus delivery. (A) In one set of rats (designated Rat 1), whisker vibration data were collected during 'electrical whisking', induced by stimulation of the exposed facial nerve. An optical sensor, shown schematically by two orthogonal light paths on the rat's snout, monitored vertical and horizontal whisker motion. (B) Closer view of the optical sensor. The sensor consisted of an LED light source and phototransistor. (C) While recording neuronal activity, whisker trajectories were played back to a second group of rats (designated Rat 2) through a piezoelectric motor, shown schematically by the horizontal and vertical arrows at the base of the whisker. (D) Piezoelectric motor. The motor was constructed from two orthogonal pairs of parallel piezoelectric wafers which were driven independently by horizontal and vertical signals. Inset gives higher magnification showing tube into which whisker was inserted. (E) Texture trajectories. Three experimental conditions are shown: free whisking (no object contact; left panel), whisking across P280 sandpaper (middle panel), whisking across P100 sandpaper (right panel). Below, the associated whisker trajectories, are shown. Each point, separated by 1 ms, gives vertical and horizontal position; the trajectory begins with protraction (unfilled right arrow) at t = 0 and terminates 125 ms later at the end of retraction (black left arrow). Whisker speed along the trajectory is given by the grey scale. The illustrated whisking trajectories come from whisker C3, and each trajectory is whisk number 201 of the 400 whisks delivered for each texture during the recording sessions.

A whisker velocity profile

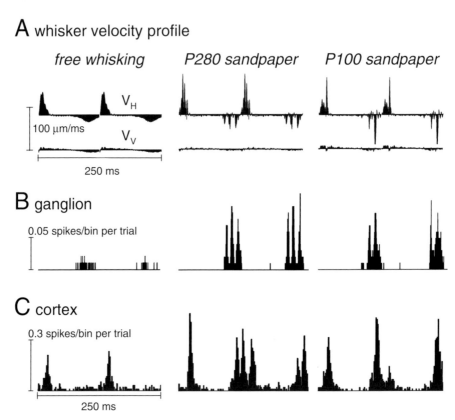

FIG. 3. Whisker velocity profiles and neuronal coding. (A) Whisker trajectories displayed according to the horizontal and vertical velocities (V_H and V_V). For V_H, whisker protraction (forward movement) is positive, whisker retraction (backward movement) negative. For V_V, upward movement is positive, downward movement negative. Each profile shows two forward and backward whisks, averaged from 400 whisks. (B) Receptor cell PSTHs (0.2 ms bins) aligned with the whisker trajectories above. (C) Cortical PSTHs (2 ms bins).

Second, to search for distinctive neuronal representations, we played back the texture-induced vibrations to the base of a whisker (method illustrated in Fig. 2C,D) while recording neuronal activity of the neurons in the trigeminal ganglion and the barrel cortex. The second set of rats thus received whisker vibrations identical to those previously recorded during active whisking in the first set of rats. The stimulus set was constructed by splicing together the trajectories associated with different textures at the point of maximum retraction where $V_H = 0$. Thus, texture stimuli were intermixed without the introduction of any position or velocity discontinuity. Peristimulus time histograms (PSTHs) from one representative receptor–cortex neuronal pair illustrates the main finding: time-varying neuronal activity in the trigeminal ganglion and cortex captured the kinetic signatures of the texture-induced vibrations. For the ganglion cell (Fig. 3B), several coding properties are evident:

- it fired a greater number of spikes for the coarse P280 and P100 textures than for free whisks
- the spikes were closely aligned to instants in which the whisker moved at high velocity
- it fired in a reproducible manner across trials—the spikes were aligned
- it was selective to whisker retraction

The barrel cortex cell cluster (Fig. 3C) was recorded simultaneously with the receptor cell, allowing direct comparison of different stations along the sensory pathway. Like the ganglion cell, the cortical cluster responded to high velocities and, as a result, fired a greater number of spikes for P280 and P100 sandpapers than for free whisks. Key differences from the receptor cell are clear:

- there was more variability in the temporal alignment of spikes
- the cluster fired for both whisker protraction and retraction

Neurons in the barrel cortex are known to encode, by the number of spikes, the mean speed of whisker vibration (Arabzadeh et al 2003, 2004). Therefore, surfaces that evoked whisker vibrations of nearly equal mean speed, such as sandpapers P280 and P100, also evoked nearly equal spike counts: the mean value of the PSTHs in Figure 3C were equivalent. The number of spikes per whisk, therefore, cannot by itself provide the neuronal basis for the discriminability between these two textures. What features of neuronal activity do permit fine texture judgements? From the illustrated data, it is clear that textures were readily discriminable by *spike patterns*. The patterns arose from the alignment of spikes to the velocity profile of the input vibration. The receptor cell reported the velocity profile for whisker retraction while the cortical cell cluster reported both protraction and retraction profiles, albeit with lower fidelity to individual velocity features. Other studied receptor cells were selective for whisker protraction: the directional non-selectivity of cortical neurons presumably resulted from the convergence of receptor cells possessing different directional selectivities.

From basic response properties to texture coding: building the bridge

We hypothesized that the texture-specific firing patterns of sensory receptors and cortical neurons could be explained by their extraction of elemental kinetic features from the whisker input signal. To test this, we presented a 10 minute 'white noise' stimulus in which two stimulus features, horizontal and vertical whisker velocity (V_H

and V$_V$), varied randomly across time. The goal was to map out how neurons encode velocity. Figure 4A illustrates the noise stimulus. Our method of 'forward correlation' between stimulus events and neuronal responses required subdividing velocity space into discrete segments; in Figure 4B, we subdivided velocity space into 8 angular (A_{1-8}) and 10 radial (R_{1-10}) sectors. We then constructed firing probabilities in relation to millions of occurrences of each velocity event, such as the velocity event A_5,R_9 (indicated by asterisk), selected as an example.

For the receptor cell (same cell as illustrated in Fig. 3), spike probabilities in the 1–2 ms post-event interval are given in Fig. 4C for all joint events (A,R). Here, velocity space has been further subdivided (20 angular and 10 radial segments) to construct the cell's 'kernel' in finer detail. The cell emitted spikes with increasing probability as velocity increased but only for restricted directions, preferring high speeds that combined retraction (backward movement) and upward movement. For the cortical cell cluster, spike probabilities in the 5–20 ms post-stimulus interval are given in Fig. 4D; like the receptor cell, the cortical cluster emitted spikes with increasing probability as velocity increased, but its directional selectivity was less pronounced and was radially symmetrical.

Responses to these noise stimuli allowed us to quantify velocity sensitivity and then simulate texture-induced spike trains based on the sequence of velocity events in the actual texture-induced vibrations. More specifically, we projected hundreds of whisk velocity trajectories upon the white noise-derived receptor and cortical kernels. One whisk on P280 sandpaper is depicted on both kernels (Fig. 4C,D). Each instantaneous velocity (A,R) during the whisk gave an ensuing spike probability profile. To simulate a spike train, a spike was generated in each time bin t with probability $P(t)$, given by the average of the overlying spike probabilities caused by the preceding velocity events. Comparison between simulated and observed responses reveals the extent to which the real responses to natural stimuli could be explained by neuronal selectivity to velocity features. Figure 5 shows the simulated receptor cell and cortical responses to two successive whisks, delivered 100 times, for free whisking and for whisking on P280 and P100 sandpapers. Comparing the PSTHs obtained from receptor and cortical simulations (overlaid dotted lines) against the real ganglion and cortical PSTHs (solid black bars, reproduced from Fig. 3), reveals a close match. The Pearson correlation coefficients between the predicted and observed PSTHs were >0.9 for the receptor cell and >0.8 for the cortical cluster. Thus, for the illustrated neurons the velocity feature extraction properties of the neurons are sufficient to explain texture responses.

In previous work done in the visual and auditory modalities, the artificial/natural 'bridging' procedure provided predicted neuronal output that failed to closely match real neuronal output (Theunissen et al 2000, 2001, Machens et al 2004). In contrast, for the selected neurons, the procedure worked well. The first reason is that we derived the kernels by mapping neuronal responses to the same feature—velocity—

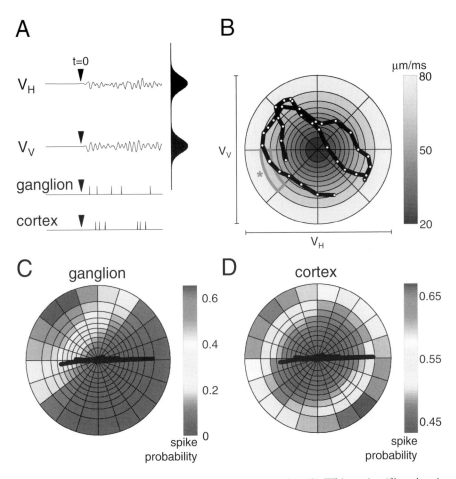

FIG. 4. Derivation of neuronal feature extraction properties. (A) White noise (filtered at 1–500 Hz) in horizontal and vertical velocities (V_H and V_V). Velocities were sampled at 0.13 ms timesteps from a Gaussian distribution with mean 0 (right side). Stimulus onset (t = 0 ms) and 40 ms of stimulation are shown, with ganglion and cortical spike trains temporally aligned. (B) 5 ms trajectory of velocity white noise. Radial coordinates give horizontal and vertical velocity (VH, VV). Velocity space was subdivided such that each segment included the same number of events (~3 435 300) across 10 minutes of stimulation; therefore, outer segments are wider. One segment (asterisk) is selected to indicate the next step of the procedure: after occurrence of each such event, spike times were accumulated to build up a spike probability profile. (C) Receptor cell spike probabilities, given by grey scale, in relation to joint VH, VV events. To estimate the kernel in finer detail, the number of angles is increased to 20. Each segment now contains about 1 374 120 velocity events. One P280 whisk trajectory is superimposed. (D) Cortical cell cluster spike probabilities, given by grey scale, in relation to joint VH, VV events. P280 whisk trajectory is superimposed.

predicted and observed neuronal responses

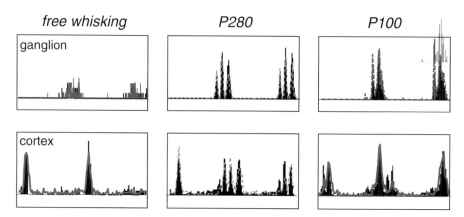

FIG. 5.   Responses to natural stimuli predicted from elemental properties. Ganglion cell and cortical cell cluster simulated (dotted line) and real (black bars) PSTHs for free whisking and two sandpapers.

which distinguishes one texture vibration from another. Moreover, our 'forward correlation' method assumes linearity and could function only if neurons were linear; that is, if ongoing response depends upon an integration process where preceding events affect the cells independently of one another. The studied neurons were, to a first approximation, linear.

Experiments using ramp-like movements (Shoykhet et al 2000) and temporally unstructured movement (Jones et al 2004) led to confusion about whether neurons encode position, velocity or acceleration. On the other hand, the use of sinusoidal stimuli showed that whisker *velocity* is the feature encoded by cortical neurons (Arabzadeh et al 2003, 2004). This raised the possibility that neurons represent texture by encoding the kinetics of whisker vibrations, but crucial questions were unanswered: does whisker movement across different textures produce distinct vibrations? If so, do neurons in the central pathway reliably report these vibrations? Through what coding mechanisms? These problems could be addressed only by:

- characterizing the natural texture stimulus, and
- applying this natural stimulus while recording neuronal activity

We have found that as the rat actively sweeps the whisker across a surface, whisker vibrations transmit the surface features to the receptors in the skin at the base of the whisker. The receptor cells, as well as cells in the somatosensory cortex, produce trains of impulses that codify the whisker vibration with remarkable precision. These impulse sequences, distinct for each vibration, are the words that form the neuronal language of textures. Texture representation is possible because the receptor cell is tuned to the very same feature of the whisker movement that best distinguishes one texture from another (the kinetic signature of the vibration). Cortical cells conserve this kinetic signature in their firing patterns.

Future directions

Although we now have taken the first steps toward understanding the neuronal basis of texture sensation, an immeasurable amount of work remains before the picture is complete! Remaining problems include the following.

The first is 'texture constancy'. Just as the visual system identifies a familiar face at different distances and under different sorts of lighting, we assume that the tactile system can identify a familiar texture under different 'whisking' conditions. Degrees of freedom include the active whisking frequency (which the rat modulates, from moment to moment, within a range from 5–20 Hz), the front–back position and radial distance of the contacted object with respect to the snout (see Derdikman et al 2005, this volume), and the set of whiskers that contacts the surface. This final point requires further explanation. Whisker length varies systematically across the anterior–posterior dimension of the rat's snout (Brecht et al 1997) and, as a consequence, the mechanical properties of whiskers vary systematically according to their location on the snout (Niemark et al 2003, Andermann et al 2004). The mechanical differences might cause different whiskers to transduce the same texture into different vibrations. At what level of the sensory system—and through what mechanisms—does neuronal activity settle upon some texture-specific 'attractor' in spite of varying input signals?

Another issue for future work is the intracortical transformation of neuronal codes. Our current data trace neuronal firing patterns between receptor neurons and the input layer of somatosensory cortex. What is the neuronal representation of texture at successive stages of cortical processing, such as secondary somatosensory cortex (SII)? In primate visual system, early cortical stages employ dense coding, whereby a large population of neurons participates in the representation of nearly every stimulus; thus, individual neurons encode many different stimuli. At later stages (e.g. inferotemporal cortex) this transforms to sparse coding, whereby a small subset of neurons participates in representing each stimulus, firing few spikes: the *occurrence* of spikes in specific neurons, more than the *patterns* of spikes among large populations, seems to be the main information-carrying feature. Our intuition is that the intracortical streams of the whisker tactile system give rise to a similar coding transformation. This could occur if late-stage neurons fire only when they receive specific patterns of early-stage neuronal output.

A final intriguing issue is the neuronal basis of the *expectation* or *prediction* of texture. As a rat learns to discriminate between two textures, across thousands of training trials, it is likely that the percept of texture arises not purely from a 'bottom-up' receptor-to-cortex pathway, but from the interaction between the 'bottom-up' signal and the expected sensation. The latter would be generated through 'top-down' processes. Decisions might arise from the match or mismatch between the expected neuronal activity pattern and the stimulus-evoked neuronal activity pattern. What might be the neuronal manifestation of top-down processes, and what is the form of the match/mismatch mechanism?

These problems are interrelated, and cannot be isolated and studied one at a time. Rather, they can be addressed only by measuring and analysing neuronal firing in awake, behaving rats, currently one area of focus of our research.

Acknowledgements

Supported by European Community IST-2000-28127, Telethon Foundation GGP02459, J.S. McDonnell Foundation 20002035, and Human Frontiers Science Programme RGP0043.

References

Adelman TL, Bialek W, Olberg RM 2003 The information content of receptive fields. Neuron 40:823–833

Ahissar E, Sosnik R, Haidarliu S 2000 Transformation from temporal to rate coding in a somatosensory thalamocortical pathway. Nature 406:302–306

Andermann ML, Ritt J, Neimark MA, Moore CI 2004 Neural correlates of vibrissa resonance: bandpass and somatotopic representation of high-frequency stimuli. Neuron 42:451–463

Arabzadeh E, Petersen RS, Diamond ME 2003 Encoding of whisker vibration by rat barrel cortex neurons: implications for texture discrimination. J Neurosci 23:9146–9154

Arabzadeh E, Panzeri S, Diamond ME 2004 Whisker vibration information carried by rat barrel cortex neurons. J Neurosci 24:6011–6020

Baylor DA, Nunn BJ, Schnapf JL 1984 The photocurrent, noise and spectral sensitivity of rods of the monkey Macaca fascicularis. J Physiol (Lond) 357:575–607

Bermejo R, Vyas A, Zeigler HP 2002 Topography of rodent whisking—I. Two-dimensional monitoring of whisker movements. Somatosens Mot Res 19:341–346

Brecht M, Preilowski B, Merzenich MM 1997 Functional architecture of the mystacial vibrissae. Behav Brain Res 84:81–97

Brown AWS, Waite PME 1974 Responses in the rat thalamus to whisker movements produced by motor nerve stimulation. J Physiol 238:387–401

Carandini M, Heeger DJ, Movshon JA 1997 Linearity and normalization in simple cells of the macaque primary visual cortex. J Neurosci 17:8621–8644

Carvell GE, Simons DJ 1990 Biometric analyses of vibrissal tactile discrimination in the rat. J Neurosci 10:2638–2648

Derdikman D, Szwed M, Bagdasarian K et al 2005 Active construction of percepts about object location. In: Percept, decision, action: bridging the gaps. Wiley, Chichester (Novartis Found Symp 270) p 4–17

Diamond ME 1995 Somatosensory thalamus of the rat. In: Jones EG, Diamond IT (eds) Cerebral cortex, Vol 11: barrel cortex. Plenum Press, New York, p 189–219

Ebara S, Kumamoto K, Matsuura T, Mazurkiewicz JE, Rice FL 2002 Similarities and differences in the innervation of mystacial vibrissal follicle-sinus complexes in the rat and cat: a confocal microscopic study. J Comp Neurol 449:103–119

Guic-Robles E, Jenkins WM, Bravo H 1992 Vibrissal roughness discrimination is barrel-cortex dependent. Behav Brain Res 48:145–152

Hutson KA, Masterton RB 1986 The sensory contribution of a single vibrissa's cortical barrel. J Neurophysiol 56:1196–1223

Jones LM, Depireux DA, Simons DJ, Keller A 2004 Robust temporal coding in the trigeminal system. Science 25:1986–1989

Kayser C, Salazar RF, Konig P 2003 Responses to natural scenes in cat V1. J Neurophysiol 90:1910–1920

Machens CK, Wehr MS, Zador AM 2004 Linearity of cortical receptive fields measured with natural sounds. J Neurosci 24:1089–1100

Neimark MA, Andermann ML, Hopfield JJ, Moore CI 2003 Vibrissa resonance as a transduction mechanism for tactile encoding. J Neurosci 23:6499–6509

Shoykhet M, Doherty D, Simons D 2000 Coding of deflection velocity and amplitude by whisker primary afferent neurons: implications for higher level processing. Somatosens Mot Res 17:171–180

Simoncelli EP, Olshausen BA 2001 Natural image statistics and neural representation. Ann Rev Neurosci 24:1193–1216

Szwed M, Bagdasarian K, Ahissar E 2003 Coding of vibrissal active touch. Neuron 40:621–630

Theunissen FE, Sen K, Dope AJ 2000 Spectral-temporal receptive fields of nonlinear auditory neurons obtained using natural sounds. J Neurosci 20:2315–2331

Theunissen FE, David SV, Singh NC, Hsu A, Vinje WE, Gallant JL 2001 Estimating spatio-temporal receptive fields of auditory and visual neurons from their responses to natural stimuli. Network 12:289–316

Thomson AM, Bannister AP 2003 Interlaminar connections in the neocortex. Cereb Cortex 13:5–14

Welker C 1971 Microelectrode delineation of the fine grain somatotopic organization of SmI cerebral neocortex in albino rat. Brain Res 26:259–275

Whitfield IC 1979 The object of the sensory cortex. Brain Behav Evol 16:129–154

Woolsey TA, Van der Loos H 1970 The structural organization of layer IV in the somatosensory region (SI) of mouse cerebral cortex. The description of a cortical field composed of discrete cytoarchitectonic units. Brain Res 17:205–242

DISCUSSION

Scott: Are there any biases in the directional tuning of the neurons? If so, is this related to the mechanics of the whiskers or the natural statistics of how these whiskers move? Is there then a shift between the biases in directional tuning between ganglia and what is seen at the cortex?

Diamond: The population of ganglion cells we have studied so far is not large enough for us to make any strong statement. We have 14 neurons. We have seen preferred directions all around the clock. So far we are not able to detect any trend. The other question is, why is there any directional selectivity? It might arise from the location of the nerve termination on the whisker shaft. The position of the termination might determine which direction of movement the cell is selective for. The cortical neurons are less selective than ganglion cells in every case. Cortical neurons usually have some preferred diagonal axis and they fire even for movements in the non-preferred direction; this is different from ganglion cells which are absolutely silent for movements in the non-preferred direction.

Logothetis: The whiskers of the rats are of different lengths. This would imply that they also have different resonance frequencies. With all this fine discrimination in space and vibrations, how do you account for different signals coming from different whiskers?

Diamond: The first systematic observation of the differences in whiskers was by Michael Brecht (Brecht et al 1997). He noted the precise relationship between the anterior–posterior position of the whisker and its length: the ones in the front are shorter. How the rat uses these differences is not known. As you say, the long whiskers have lower resonance frequencies; the shorter ones have higher resonance frequencies. As a result, one laboratory is pushing the idea that the rat uses resonance frequencies to detect textures. The idea would be that texture with a low spatial frequency—grains or ridges far apart from each other—would produce a low temporal frequency as the rat moves the whisker across. This would cause the whiskers with low resonance frequencies to resonate. A surface with high spatial frequencies would make the shorter whiskers resonate. Then the rat would use its cortical map of position to find out which whiskers resonate. However, our evidence doesn't support this hypothesis. When we study whiskers they seem to be highly damped in the whisker follicle. Also, the resonance frequency depends strongly on the boundary conditions, such as how hard the whisker is pushing against the surface. In behaving rats there is a lot of variability in this, such as variability in the length of the free part of the whisker compared with the part in contact. How much is free in the air would strongly affect resonance frequency. For the rat to use resonance frequencies it might have to take into account the changes in resonance frequency produced by the contact itself. This seems like a complicated computation. We think the computation is more likely a measurement of the temporal profile of vibration.

Logothetis: You had a correlation coefficient of 0.8 between the model and the cortex. This would imply that you have close to 35% unexplained variance. Would this 35% unexplained variance be accounted for by looking into the input you get from each one of the whiskers, if they have different resonance frequencies?

Diamond: I don't think so, because we find a high performance of the output of the model even when the vibration picked up from one whisker is applied to a different whisker. One neuron seems to fire in the way it should even for another whisker's vibration. The correlation is a bit of a conservative one, in the sense that if you divide the real trials in half, and measure the correlations between the histograms of the two halves of the data, it is about 85% for cortical neurons. Some of the non-explained variability is just because real cortical neurons vary across stimulus repetitions. Thus, the simulated data are more than 90% as good as they could possibly be (Arabzadeh et al 2005).

Logothetis: You showed that the responses in the ganglion cells were only to the negative movement and not to the positive. Is this correct?

Diamond: No, some ganglion cells prefer one direction and others another direction.

Wolpert: In the previous paper we heard that curvature is important and angle seems to be relatively unimportant. I understand that your motor-driven system will change the angle, but not change curvature. Is this a problem?

Diamond: This still has to be sorted out. When we used this method of playing back a vibration we are in fact not applying curvatures. There could be additional information that is coming through the whisker that our type of experiment doesn't capture. For example, very rough surfaces that grip the whisker could produce more drag and more curvature. This isn't captured by the playback method. We chose to emphasize the fact that while there may be more information about the texture that is available, even the simple vibration by itself without curvature or tension still carries a lot of information.

Wolpert: Why is texture important to rats? For us, I can understand that if we want to pick something up we need to know how rough the surfaces is so we know how hard to squeeze.

Diamond: I can imagine that rats would find their way about in their environment, for example a dark basement, according to the walls. They may know the path from one place to another according to the different surfaces of the objects that they touch.

Albright: Is whisking frequency completely independent of the velocity with which the whisker is moved across the surface? If the animal whisks at a higher rate, will it be moving faster?

Diamond: My impression is that when rats whisk at a higher frequency, the arc of the whisker movement decreases. They go through less of a range.

Albright: So the kinetic signature is going to be the same, independent of whisking frequency; what I am concerned about is that if the velocity changes then a given texture might have a different kinetic signature.

Diamond: You sound like the reviewer for my last paper! What you are talking about is an important and general problem: how can the rat create constancy even when they change the conditions? We don't know this at all, and we need to study it more. There is one interesting observation, which comes from Carvell & Simons (1990), where they trained rats to make a difficult tactile discrimination. In this experiment, they observed that over the course of the training the movements the rats made became increasingly similar over time. They assumed a particular posture and moved the whiskers over a particular range with a particular frequency. In fact, rats that showed less variability in whisking across trials performed better. The rat may try to reduce some of that variability by controlling its movements.

Barash: Is it clear that there is no torsion? You have horizontal and vertical movements but could there also be information from torsion? Also, are the cells that you described what Dori Derdikman described as 'touch cells'?

Diamond: We didn't classify the cells using the same scheme. One project our two laboratories want to do together is to measure neuronal responses both when rats are doing the localization task and when they are assessing texture, to see whether a cell can accomplish both tasks. Are cells selectively involved in localization coding or texture coding? At this point we don't know.

Barash: Would you expect it to be one class or distributed across classes?

Diamond: I don't feel confident enough to make a prediction.

Barash: You described elegant time locking of the stimulus, which becomes less precise in the cortex. What would be the implications for information processing? How could this time locking, which is then being loosened, be specifically important for let's say surface recognition?

Diamond: There are two kinds of codes that the system seems to be using for textures. One, which I didn't mention, is the quantity of spikes that the neurons emit. It turns out that when textures produce more energetic vibrations, the neurons fire more spikes. Perhaps this is not surprising. If we simply count the number of spikes we have some degree of information that distinguishes one texture from another. But some textures produce vibrations of the same mean energy and therefore the same spike count. In this case there wouldn't be adequate information from counting spikes. These textures, however, produce very different temporal patterns. We assume that the sensory system makes use of the temporal patterns in decoding the textures.

Rizzolatti: The cortical responses seem to be variable and rather ambiguous. Does the cortex use a different code to recognize texture with respect to the ganglion cells?

Diamond: We can still recognize the texture according to the cortical firing pattern, but it is more variable across trials. The ganglion cell has only one input, its principal whisker. All it can report is this whisker, but cortical neurons have many inputs. Their activity reflects their position within a very complicated network. One interesting feature of the cortex, which we are not sure how to interpret, is that the trial–trial variability on one texture for the ganglion cell is large, because the tiny details of the movement vary even when the whisker moves across the same texture in different whisks. For the cortical cells, the trial-to-trial variability on the same texture is no larger than the trial-to-trial variability when one exact stimulus is replayed. Therefore the cortex is in a sense making the response fit into a particular category and is ignoring the differences that occur on each trial. It seems like it is perhaps cancelling a bit of the sensory noise.

Brecht: Then why do you need such a huge time resolution in the cortex? This is never seen in the visual system.

Logothetis: It would suggest a system sensitive to phase.

Brecht: The visual system is slow because the retina cannot do better. It comes down to the channels, and the mechano-gated channels are much faster than the G-coupled channels in the retina.

Logothetis: If you go with the valley principle, you take these times and extract space from oscillations much higher than the frequency of the fastest cell ever firing. If you want to go to 22 kHz and you have a neuron that can fire at a maximum frequency of 1 kHz, you have to rely on some kinds of different encoding principles.

Gold: Do you think it is possible to gain insight into these issues? Is this kind of cortical noise important in the context of the kind of behavioural constancy that you were talking about before, where rats trained on difficult tasks tend to use stereotyped movements? Can you do a perceptual discrimination where you minimize variability in behaviour and then compare variability in perceptual choices of the whole animal compared with the variability in cortex? Is the variability of perception more comparable to what is seen in the nerve, or what is seen in barrel cortex? Has anyone done this?

Diamond: No one has done this in the whisker system. It is something that would have to be studied in the behaving rats, which is why we want to do the recordings from behaving rats. The variability I showed you was in anaesthetized rats. Rhythmic activity related to anaesthesia may be part of the variability. In the awake rats it is likely that the system is much quieter.

Sparks: If I understand correctly, you could have different whisk rates and texture patterns that would give you the same pattern of receptor output. If the animal could discriminate between those, this would be strong evidence that the motor pattern was being used to decode this information. Do we know anything about the ambiguity that comes from a combination of whisk rate and texture pattern?

Diamond: I don't think we know much about it. We have indirect evidence only: as rats are trained on a difficult discrimination they tend to take a stereotyped whisking approach to the surface each time. Within the whisking range of 5–20 Hz they narrow in on a small part and apply that every time. It seems as if they want to reduce the motor variability. There is a very strong motor pattern generator in the brain stem, and no one is certain how this feeds into the sensory system. It would be surprising if the rat didn't use its own motor knowledge to interpret sensory information.

Krubitzer: It seems like the rat is actively interfacing its whisker system with the object. Has anyone looked simultaneously at the sort of things you are looking at and the proprioceptors of the head and neck? It seems like arc of the head would tell you something about the space that the animal is moving in as well as the size of the object. It would have to be a different arc depending on the size of the object, for instance. If the animal is using only the whiskers to gather information about the world, it might explain some of the variability, because they are presumably also getting inputs from the proprioceptors in the head, neck and face.

Diamond: At the very least, the rat's own movement has to fit into the equation of the motor system. In this task, a lot of the movement comes from the rat moving its head. Beyond the rat knowing how it is moving its own whiskers, which is already complicated, it likely also knows how fast it is walking and how it is turning its head. All these will affect the whisker movement across a surface.

Krubitzer: If you had some kind of mutant mouse with motor difficulties, you might expect differences in the recordings you are getting in cortex, for example.

Haggard: I am also interested that the variability goes up as you go into the cortex. I am struck by the fact that in your experimental model there is only a single whisker being examined. If you have multiple whiskers, then the cortical signals could be integrated, perhaps by some unknown stage or secondary process. When you integrate two signals which each have a bit of temporal noise, the effective temporal resolution could be improved. Is anything known about how the cortex integrates information coming from neurons representing several whiskers?

Diamond: I would be happy to hear from Michael Brecht and Dori Derdikman about that. This is an area people have been arguing about for 20 years. There are two extreme positions. One is that an individual cortical neuron is reporting only its own whisker, and then the opposite point of view—that one function of cortex is to bring together the whiskers. It is not clearly resolved, but this is one of the many questions that needs more behavioural work. You can imagine that under some circumstances the rat would want to segregate the information from different whiskers, for example for a localization task. In other circumstances it may be important for the rat to combine information from different whiskers. My guess is that the pathway can shift among different sorts of integration according to the behavioural task.

Haggard: This makes a lot of sense, particularly in the context of texture, where temporal resolution is important. If you could combine a number of signals you might have a relatively high temporal resolution.

Derdikman: At a certain stage (probably higher than S1), what interests the rat is where the object is much more than where its whiskers are. At least somewhere along the processing pathway, the system of coordinates must change. For the rat, the question is not which whisker is touching, or what is the receptive field of these specific neurons, but what is the position in space of the object the whiskers were touching.

Harris: Have you observed a difference in the amount of time a rat will spend palpating a positive stimulus versus a negative stimulus?

Diamond: In our behavioural experiments, where rats palpate a rough or smooth surface as a cue for food location, our impression is that they spend more time touching the negative stimulus, because they have to make an active decision of stopping and turning around. They have to inhibit their natural tendency to run into each arm.

Romo: Can you speculate on the texture coding processing scheme in the rat somatosensory cortices?

Diamond: My guess is that the coding sparses, such that neurons become selective for specific textures. In primary sensory cortex we see that every neuron encodes every texture, by spike–count and by firing patterns. The speculation would be that in S2 or other cortical areas, not every neuron is involved in representing every texture, but rather whether or not the neuron fires spikes actually carries information about the texture.

Romo: Would you agree that the primary somatosensory cortex alone is not sufficient to achieve discrimination? Or do you think this is enough together with the motor cortex?

Diamond: It depends on what you mean by 'sufficient to achieve': I think all the information the rat uses to make the discrimination is in the primary somatosensory cotex.

References

Arabzadeh E, Zorzin E, Diamond ME 2005 Neuronal encoding of texture in the whisker sensory pathway. PLoS Biol 3:e17

Brecht M, Preilowski B, Merzenich MM 1997 Functional architecture of the mystacial vibrissae. Behav Brain Res 84:81–97

Carvell GE, Simons DJ 1990 Biometric analyses of vibrissal tactile discrimination in the rat. J Neurosci 10:2638–2648

Cortical commands in active touch

Michael Brecht

Erasmus MC, University Medical Center Rotterdam, Department of Neuroscience, Dr. Molewaterplein 50, 3015 Dr Rotterdam, The Netherlands

Abstract. The neocortex is an enormous network of extensively interconnected neurons. It has become clear that the computations performed by individual cortical neurons will critically depend on the quantitative composition of cortical activity. Here we discuss quantitative aspects of cortical activity and modes of cortical processing in the context of rodent active touch. Through *in vivo* whole-cell recordings one observes widespread subthreshold and very sparse evoked action potential (AP) activity in the somatosensory cortex both for passive whisker deflection in anaesthetized animals and during active whisker movements in awake animals. Neurons of the somatosensory cortex become either suppressed during whisking or activated by an efference copy of whisker movement signal that depolarize cells at certain phases of the whisking cycle. To probe the read out of cortical motor commands we applied intracellular stimulation in rat whisker motor cortex. We find that APs in individual cortical neurons can evoke long sequences of small whisker movements. The capacity of an individual neuron to evoke movements is most astonishing given the large number of neurons in whisker motor cortex. Thus, few cortical APs may suffice to control motor behaviour and such APs can be translated into action with the utmost precision. We conclude that there is very widespread subthreshold cortical activity and very sparse, highly specific cortical AP activity.

2005 Percept, decision, action: bridging the gaps. Wiley, Chichester (Novartis Foundation Symposium 270) p 38–50

Rodent active touch as a model system

The brain processes and represents sensory and motor information in tightly interconnected and largely overlapping neural circuits. This is beautifully documented for the rodent vibrissal system, which can be conceptualized as a hierarchy of sensorimotor loops. Here, sensorimotor signalling occurs at all levels starting with connections between brainstem sensory and motor neurons, up to high-level sensorimotor interactions between cortical areas and cortico-cerebellar interactions (Kleinfeld et al 1999). While the great functional significance of the tight sensorimotor link is largely undisputed, the combined study of sensory and motor processing is the exception rather than the rule. Here I discuss issues of sensorimotor processing in cortical circuits of the rodent active touch system (Carvell & Simons 1990, Krupa et al 2004). This is a superb model system, where sensation (whisker contacts) and action (whisker movements) form a functional entity (active touch);

in addition, the rodent whisker system offers a large number of experimental advantages (ease of manipulation of whiskers, exquisitely organized neural representations—barrels, barreloids, barrelettes, etc.—of whisker information, Jones & Diamond 1995). Attracted by such advantages experimentalists have applied a range of novel techniques to study barrel cortex circuits and function (Brecht et al 2004a). In this review I will outline how novel techniques, in particular intracellular stimulation experiments and whole-cell recordings in anaesthetized and behaving animals have altered our ideas about the representation of active touch and cortical processing on the whole. The sensorimotor nature of active touch will be of decisive importance in this analysis, as it allows one not only to study how sensory representations are formed, but also to probe the read out of cortical activity.

Cortical cells, cortical activity and modes of cortical processing

Cortical neurons form an enormous network that constitutes the largest part of the mammalian brain. Thus, cortical neurons greatly outnumber peripheral sensory and motor neurons. In the rat vibrissae/barrel cortex system we find about 200 primary afferents per whisker, whereas estimates based on cell density (Keller & Carlson 1999, Gottlieb & Keller 1997) suggest that on the order of 20 000 cortical neurons are situated in the corresponding barrel column. A similar calculus holds for the motor side of the system, where in the order of 30 000 motor cortical neurons converge directly (Grinevich et al 2004) and indirectly (Hattox et al 2002, 2003) on a small number of facial nucleus vibrissa motoneurons—in the order of 50–100 cells per whisker (Klein & Rhoades 1985).

Since the complete sensory information flows through the primary afferents and all whisker movements are mediated by the facial nucleus cells, one cannot help wondering why there are so many cortical neurons. Indeed, interpreting the large number of cortical neurons has been a key question in cortical neurobiology, but as yet no unifying answer to the problem has emerged. What has become clear, however, is that understanding the significance of individual or numerous cortical neurons is closely linked to developing models of cortical activity (Shadlen & Newsome 1998).

Cortical neurons are extensively interconnected. For example a layer 2/3 neuron in rat cortex may receive and send about 10 000 synaptic terminals (DeFelipe & Farinas 1992). Since neurons connections between neurons often consist of around five terminals (Markram et al 1997, Feldmeyer et al 1999), each neuron will form connections with a few thousand pre- and postsynaptic cells.

Thus, if one assumes that these cells show substantial levels of action potential (AP) activity (say > 1 Hz), one would conclude that each neuron is exposed to thousands of inputs each second. In such synaptic bombardment scenarios (Desthexe & Pare 1999), the individual APs and the single neuron play a rather minor role.

Cortical processing emerges from mass action; cells compute under a so called 'high input regime' where large numbers of excitatory and inhibitory inputs cancel each other out to generate a small net response (Shadlen & Newsome 1998). Another characteristic of such scenarios is that the timing of individual APs plays only a minor role in cortical computation. As postsynaptic potentials (PSPs) fall like rain on cortical dendritic trees, their fine timing becomes indiscernible on the post-synaptic level (Shadlen & Newsome 1998). Simply as a result of the boundary conditions in such scenarios, it appears likely that cortical neurons perform their calculations in a noisy computational environment. Accordingly, the variability of cortical AP discharges is then thought to reflect noise (Shadlen & Newsome 1994, 1998). A major computation that cortical neurons perform in high input regimes is averaging. This reasoning could explain how the cortex might work with high precision despite noisy individual processing elements and ultimately, why so many cortical neurons are needed.

In this chapter I will outline an alternative view of cortical processing and I will develop these ideas discussing the example of rat active touch. In brief, I will suggest that cortical responses are generated by a few carefully selected inputs, under a so-called selective input regime (Brecht & Sakmann 2002b). Accordingly, cortical subthreshold responses may reflect the superposition of relatively few (in the order of tens), precisely timed unitary inputs. Under such conditions the precise timing of synaptic inputs determines postsynaptic output (Abeles 1983, Zador 1998). As a rule, most cortical cells remain well below the AP initiation threshold and only a few specific and precisely timed combinations of inputs generate the large excitatory postsynaptic potentials (EPSPs) that elicit one or more APs.

The representation of sensory and motor information in the rat somatosensory cortex

Synaptic transmission (Feldmeyer et al 1999, 2002, Markram et al 1997, Cowan & Stricker 2004) and postsynaptic sensory responses (Moore & Nelson 1998, Zhu & Connors 1999, Brecht & Sakmann 2002, Brecht et al 2003, Manns et al 2004) in the rat barrel cortex have been investigated in great detail. The barrel cortex is there-fore an attractive system to develop quantitative concepts of synaptic information flow and cortical processing. I will discuss data from two complementary experi-mental approaches that may help us to understand the quantitative composition of synaptic signals that set up sensorimotor representations in the somatosensory cortex.

Responses to passive whisker deflection: insights from receptive field analysis

A classic approach to study neurons in sensory cortices is to apply stimulus sets and to quantify the resulting sensory responses. The barrel cortex of anaesthetized rodents is uniquely suited for such a mapping of the stimulus space, because highly controlled single- and multi-whisker deflections can easily be applied (Simons 1983).

One remarkable finding from receptive field (RF) analysis is that almost all neurons in barrel cortex (>90% of cells) show subthreshold responses when strong stimuli to the appropriate whisker are applied. Strong stimuli (i.e. fast 6° deflections) lead to widespread cortical subthreshold activity, for example in the superficial cortical layer 2/3 about 40 000 cells show responses of ≥5 mV (Brecht et al 2003). There is remarkably good experimental agreement on the size of postsynaptic responses in the barrel cortex of anaesthetized rats. Thus, the PSP amplitudes of sensory responses recorded with sharp microelectrodes (Carvell & Simons 1988, Stern et al 2001) and whole-cell recordings (Moore & Nelson 1998, Zhu & Connors 1999, Brecht & Sakmann 2002b, Brecht et al 2003) usually differ by a factor of less than two when the average of the population of recorded cells is considered. Moreover, signals from simultaneous intracellular and local field potential recordings are well correlated (M. Brecht, unpublished observations), and simultaneous whole-cell recordings and voltage-sensitive-dye recordings show an excellent correlation (Petersen et al 2003).

At the same time the sensory evoked AP activity in barrel cortex remains controversial and seems to vary greatly as a function of the methodology employed. Thus, for layer 4 (L4) barrel cortex neurons our whole-cell recordings report a mean of 0.14 APs per principal whisker (PW; the most effective whisker for that neuron) deflection (Brecht & Sakmann 2002), a response that is 5–10-fold smaller than what has been reported for L4 neurons by extracellular single-unit recordings under the same anaesthesia (Diamond et al 1993, Armstrong-James 1995). In L2/3 we observed evoked AP rates of only 0.031 APs per PW stimulus (Brecht et al 2003), which is about 40-fold less than what has been reported by unit recordings (Armstrong-James 1995). As can be seen in Figure 1 these firing rates are indeed very low, with only a minority of cells firing spikes to powerful principal whisker stimuli. Other studies document a similar difference of low AP counts reported by whole-cell recording studies (Moore & Nelson 1998, Zhu & Connors 1999) and high AP counts reported by unit recording studies (Simons 1985, Simons & Carvell 1989).

Evidence from a variety of approaches and wide range of control experiments (Margrie et al 2002, Brecht et al 2004b) argue that low AP counts documented by whole-cell recordings may be more representative than the high firing rates found with unit recordings. The issue requires further investigation, but it is likely that unit recordings are distorted by a sampling bias against cells with low firing rates. Indeed, if the firing rate estimates of whole-cell recordings for barrel cortex cells are correct (see Fig. 1), many cells could not possibly be detected by unit recordings, because they do not fire APs. Thus the picture that emerges from RF analysis in anaesthetized animals is one of very widespread subthreshold activity after strong stimuli (involving up to 40 000 cells in cortical L2/3 alone), and rather

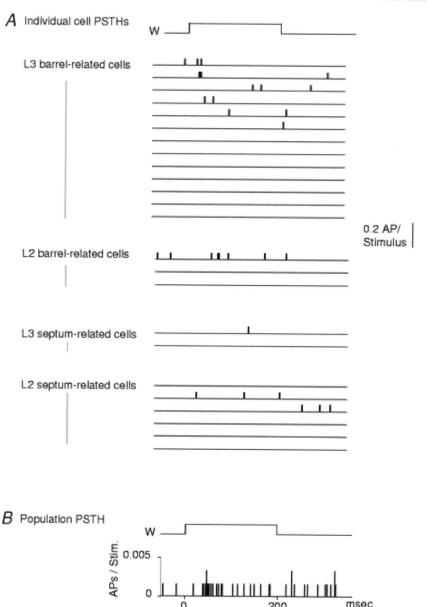

FIG. 1. Peri-stimulus time histograms (PSTHs) of sensory responses of identified supragranular pyramidal cells recorded in anesthetized rats. (A) PSTHs of responses to (6 degree) principal whisker deflection in 13 layer 3 (L3) barrel-related cells, 3 L2 barrel-related cells, 2 L3 septum-related cells, 6 L2 septum-related cells. Bin width of PSTHs is 0.5 ms. Time course of PW whisker deflection is shown above the PSTH. Data are based 20 stimulus repetitions. (B) Population PSTHs of all cells ($n = 30$). Time course of PW whisker deflection is shown above the PSTH. Modified from Brecht et al (2003).

minimal AP activity (involving only 200–300 cells with AP activity in cortical L2/3).

Responses to active whisker movement: the representation of active touch

The aforementioned data come from anaesthetized preparations. However, the more relevant question seems to be what AP activity levels are in the working brain. To address this problem we investigated cortical activity with whole-cell recordings in awake rodents. In a small number of cells we have mapped RFs in awake animals (Margrie et al 2002, M. Brecht, B. Sakmann, unpublished data). Here, we observed higher rates of sensory evoked APs than in anaesthetized animals, but these firing rates were still severalfold lower than what has been reported in unit recordings in anaesthetized animals.

Another line of investigation on AP activity in awake rats was directed to the analysis of cortical activity during active whisker movement. Acquisition of sensory information is an active process that often involves exploratory movements of eyes, ears, fingers or feelers. The active movement of sensors seems to be highly advantageous as it evolved independently in multiple sensory modalities. During rodent active touch, large rhythmic high-speed movements of whiskers ('whisking') cause ambiguities in the position of a whisker when it encounters an object. We recorded activity of L2/3 neurons in the somatosensory/barrel cortex of awake animals. Again, we find that AP activity is sparse, whereas the subthreshold membrane potential is strongly modulated during whisking. The peri-stimulus time histogram (PSTH) of whisking related activity shown in Figure 2 illustrates this point.

In one third of neurons the PSP fluctuations become 'movement suppressed' during whisking, and the remaining fluctuations in membrane potential are not correlated with whisker movements. In another third of neurons, membrane potential fluctuations are 'movement-modulated' during whisking and phase-locked to whisker movements. The remaining third of neurons show a mixture of these two response patterns. The pattern of movement-modulation of membrane potential differs between cells but is consistent within cells across whisking episodes. Specifically, different cells are depolarized at different points in the whisking cycle. Some being maximally depolarized at the beginning of whisker protraction, while others are depolarized at the middle or end of the whisker protraction; few cells were depolarized during whisker retraction.

Movement-modulated neurons receive strong synaptic input preceding the onset of movement. As can be seen in Figure 2 cells tend to discharge just before the onset of whisking, a suprathreshold reflection of the premovement inputs. These premovement inputs 'predict' both the magnitude and the pattern of movement-modulation and suggest that movement-modulation originates from an efference copy rather than from sensory inputs. Movement-modulated cells presumably dom-

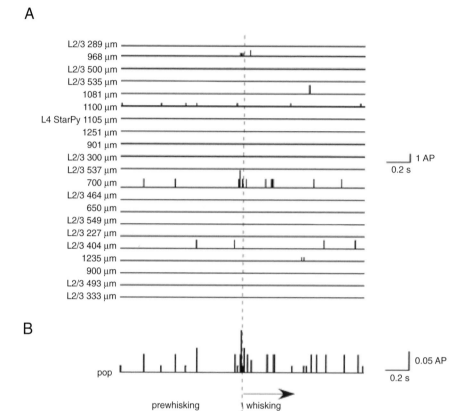

FIG. 2.   Peri-whisking time histograms (PWTHs) of AP activity in the barrel cortex of awake rats during whisking. (A) PWTHs of AP activity of regular spiking cells in the barrel cortex of awake rats one second before and one second during whisking. (B) Population PSTH. Data refer to 42 whisking episodes with a duration of ≥1 s and a well-defined beginning (an at least three-fold increase of movement amplitude at the onset of the episode). In A the depth and the putative layer of cells is given, whereby we assumed that cells at a depth <600 μm belong to L2/3. Only one L4 barrel star pyramidal neuron could be histologically recovered.

inate the ensemble excitation of somatosensory cortex during whisking and stand out from the background of movement-suppressed neurons. Thus, there seems to be a representation of motor commands in a subpopulation of neurons of the somatosensory cortex. We speculate that such an internal representation of motor commands could be used to tune neuronal responses to specific parts of the whisking cycle.

Motor commands from sparse cortical activity

The data discussed so far provide correlative evidence for sparse coding in the rodent cortex. While such measurements of cortical activity traditionally figure most prominently in the field of cortical physiology, it is also clear purely correlative research approaches by themselves are incomplete. For probing cortical codes more directly, stimulation-based approaches can provide direct insights into how cortical activity sets up neural representations (Cohen & Newsome 2004).

We used a stimulation-based approach to study cortical commands for whisker movements in the rat primary vibrissa motor cortex. Both extracellular stimulation techniques and extracellular single cell recordings have demonstrated that the activity of neurons in the primary motor cortex is closely associated with movement generation in mammals. However, neither of these techniques identifies the stimulated/recorded neurons and thus both techniques are not suited to pinpoint the motor commands issued by individual cells. We overcame this problem by applying intracellular stimulation to identified neurons in the deep layers of whisker motor cortex of lightly anaesthetized rats (Brecht et al 2004c). We found that AP initiation in individual cortical cells can cause long sequences of small whisker movements (Fig. 3).

Intracellular stimulation in layer 5 evokes movements that are phase locked from trial to trial (Fig. 3), whereas APs injected in L6 cells evoke bursts of whisking without specifying the phase of the individual movement. These observations make it plausible that L5 neurons of the primary vibrissa motor cortex are the source for providing the efference copy we observed in neurons of the somatosensory cortex.

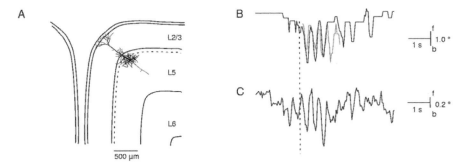

FIG. 3.   Whisker movements evoked by intracellular stimulation of single L5 and L6 cells. (A) Topographic position, and morphology of the stimulated L5 neuron. Extracellular stimulation close to the recording site evoked backward movement of a single whisker (D1). (B) Position of whisker D1 during a partially shown (grey) and a fully shown (black) intracellular stimulation trial. Note that the evoked movements are in phase. Stimulation evoked 10 action potentials (APs) at 50 Hz. (C) Movement average of 15 trials of intracellular stimulation. Modified from Brecht et al (2004c).

This reasoning is in line with anatomical studies (Veinante & Deschenes 2003) and a number of studies from Kleinfeld and colleagues, which suggest that vibrissa motor cortex can act as a pattern generator for whisking movements. These studies demonstrated a covariation of motor cortical activity and whisking (Ahrens & Kleinfeld 2004), whisking evoked by intracortical stimulation (Berg & Kleinfeld 2003) and sensory tuning for whisking frequencies in motor cortex (Kleinfeld et al 2002). These results together with the anatomical findings support the idea that the vibrissa motor cortex can act as a rhythm generator for whisking and at the same time might send an efference copy of such commands to the somatosensory cortex.

The ability of a single motor cortex neuron to trigger movements is astonishing given the large number of neurons in vibrissa motor cortex (approximately one million according to a rough estimate, Brecht et al 2004c). The brain's capacity to translate APs injected into individual cortical neurons, into noticeable motor outputs suggests that cortical APs are processed with the utmost precision. All in all these findings from intracellular stimulation appear to be incompatible with a cortical 'high input regime', where noisy cellular elements compute rough averages from a bombardment of synaptic inputs. Instead the data seem to suggest that a few spikes of an individual cell can significantly affect the animal's motor output.

Conclusion

The picture of cortical processing that emerges from our work on rodent sensori-motor representations is one of sparse AP activity and very widespread cortical subthreshold activity. How could such a dichotomy between cortical APs and PSPs be sustained? We reason that the structural basis for these observations consists of the rich synaptic connectivity of cortical neurons, where a single AP will lead to noticeable excitation in thousands of postsynaptic cells. This rich and powerful cortical synaptic transmission is constantly sculpted by synaptic learning rules. We speculate that such learning rules constrain cortical activity such that a large number of cortical neurons compete to generate a small number of APs. Thus a minimum number of cortical APs may suffice to represent sensory stimuli and generate both motor outputs as well as internal representations of these motor commands in the form of an efference copy.

References

Abeles M 1983 Role of the cortical neuron: integrator or coincidence detector? Isr J Med Sci 18:83–92
Ahrens KF, Kleinfeld D 2004 Current flow in vibrissa motor cortex can phase-lock with exploratory rhythmic whisking in rat. J Neurophysiol 92:1700–1707
Armstrong-James M 1995 The nature and plasticity of sensory processing within adult rat barrel cortex. In: Jones EG, Diamond IT (eds) The barrel cortex of rodents. Plenum Press, New York, p 333–374

Berg RW, Kleinfeld D 2003 Vibrissa movement elicited by rhythmic electrical microstimulation to motor cortex in the aroused rat mimics exploratory whisking. J Neurophysiol 90:2950–2963

Brecht M, Sakmann B 2002 Dynamic representation of whisker deflection by postsynaptic potentials in morphologically reconstructed spiny stellate and pyramidal cells in the barrels and septa of layer 4 in rat somatosensory cortex. J Physiol 543:49–70

Brecht M, Roth A, Sakmann B 2003 Dynamic receptive fields of reconstructed pyramidal cells in layers 3 and 2 of rat somatosensory cortex. J Physiol 553:243–265

Brecht M, Fee MS, Garaschuk O et al 2004a Novel approaches to monitor and manipulate single neurons in vivo J Neurosci 24:9223–9227

Brecht M, Schneider M, Manns ID 2005 Silent neurons in sensorimotor cortices: implications for cortical plasticity. In: Ebner F (ed) Neural plasticity in the adult somatic sensory-motor systems. CRC Press, Boca Raton, USA, p 1–19

Brecht M, Schneider M, Sakmann B, Margrie TW 2004c Whisker movements evoked by stimulation of single pyramidal cells in rat motor cortex. Nature 427:704–710

Carvell GE, Simons DJ 1988 Membrane potential changes in rat SmI cortical neurons evoked by controlled stimulation of mystacial vibrissae. Brain Res 448:186–191

Carvell GE, Simons DJ 1990 Biometric analyses of vibrissal tactile discrimination in the rat. J Neurosci 10:2638–2648

Cohen MR, Newsome WT 2004 What electrical microstimulation has revealed about the neural basis of cognition. Curr Opin Neurobiol 14:169–177

Cowan AI, Stricker C 2004 Functional connectivity in layer IV local excitatroy circuits of rat somatosensory cortex. J Neurophysiol 92:2137–2150

DeFelipe J, Farinas I 1992 The pyramidal neuron of the cerebral cortex: morphological and chemical characteristics of the synaptic inputs. Prog Neurobiol 39:563–607

Destexhe A, Pare D 1999 Impact of network activity on the integrative properties of neocortical pyramidal neurons in vivo. J Neurophysiol 81:1531–1547

Diamond ME, Armstrong-James M, Ebner FF 1993 Experience-dependent plasticity in adult rat barrel cortex. Proc Natl Acad Sci USA 90:2082–2086

Feldmeyer D, Egger V, Lübke J, Sakmann B 1999 Reliable synaptic connections between pairs of excitatory layer 4 neurones within a single 'barrel' of developing rat somatosensory cortex. J Physiol 521:169–190

Feldmeyer D, Lübke J, Silver RA, Sakmann B 2002 Synaptic connections between layer 4 spiny neurone-layer 2/3 pyramidal cell pairs in juvenile rat barrel cortex: physiology and anatomy of interlaminar signalling within a cortical column. J Physiol 538:803–822

Grinevich VV, Brecht M, Seeburg PH, Osten PA 2004 Direct input from vibrissa motor cortex to the rat facial nucleus revealed by lentivirus based GFP expression. Society for Neuroscience 2004, Program No. 857.22

Gottlieb JP, Keller A 1997 Intrinsic circuitry and physiological properties of pyramidal neurons in rat barrel cortex. Exp Brain Res 115:47–60

Hattox AM, Priest CA, Keller A 2002 Functional circuitry involved in the regulation of whisker movements. J Comp Neurol 442:266–276

Hattox AM, Li Y, Keller A 2003 Serotonin regulates rhythmic whisking. Neuron 39:343–352

Jones EG, Diamond IT (eds) 1995 The barrel cortex of rodents. Plenum Press, New York

Keller A, Carlson, GC 1999 Neonatal whisker clipping alters intracortical, but not thalamocortical projections, in rat barrel cortex. J Comp Neurol 412:83–94

Klein BG, Rhoades RW 1985 Representation of whisker follicle intrinsic musculature in the facial motor nucleus of the rat. J Comp Neurol 232:55–69

Kleinfeld D, Berg RW, O'Connor SM 1999 Anatomical loops and their electrical dynamics in relation to whisking by rat. Somatosens Mot Res 16:69–88

Kleinfeld D, Sachdev RNS, Merchant LM, Jarvis MR, Ebner FF 2002 Adaptive filtering of vibrissa input in motor cortex of rat. Neuron 34:1021–1034

Krupa DJ, Wiest MC, Shuler MG, Laubach M, Nicolelis MA 2004 Layer-specific somatosensory cortical activation during active tactile discrimination. Science 304:1989–1992

Manns ID, Sakmann B, Brecht M 2004 Sub- and suprathreshold receptive field properties of pyramidal neurons in layers 5A and 5B of rat somatosensory barrel cortex. J Physiol 556:601–622

Margrie TW, Brecht M, Sakmann B 2002 In vivo, low resistance, whole-cell recordings from neurons in the awake and anaesthetized mammalian brain. Pflügers Arch 444:491–498

Markram H, Lubke J, Frotscher M, Roth A, Sakmann B 1997 Physiology and anatomy of synaptic connections between thick tufted pyramidal neurones in the developing rat neocortex. J Physiol 500:409–440

Moore, CI, Nelson, SB 1998 Spatio-temporal subthreshold receptive fields in the vibrissa representation of rat primary somatosensory cortex. J Neurophysiol 80:2882–2892

Petersen CC, Grinvald A, Sakmann B 2003 Spatiotemporal dynamics of sensory responses in layer 2/3 of rat barrel cortex measured in vivo by voltage-sensitive dye imaging combined with whole-cell voltage recordings and neuron reconstructions. J Neurosci 2003 23:1298–1309 Erratum in: J Neurosci 2004 24:1 p following 906

Shadlen MN, Newsome WT 1994 Noise, neural codes and cortical organization. Curr Opin Neurobiol 4:569–579

Shadlen MN, Newsome WT 1998 The variable discharge of cortical neurons: implications for connectivity, computation, and information coding. J Neurosci 18: 3870–3896

Simons DJ 1983 Multi-whisker stimulation and its effects on vibrissa units in rat SmI barrel cortex. Brain Res 276:178–182

Simons DJ 1985 Temporal and spatial integration in the rat SI vibrissa cortex. J Neurophysiol 54:615–635

Simons DJ, Carvell GE 1989 Thalamocortical response transformations in the rat vibrissa/barrel system. J Neurophysiol 61:311–330

Stern EA, Maravall M, Svoboda K 2001 Rapid development and plasticity of layer 2/3 maps in rat barrel cortex in vivo. Neuron 31:305–315

Veinante P, Deschenes M 2003 Single-cell study of motor cortex projections to the barrel field in rats. J Comp Neurol 464:98–103

Zador A 1998 Impact of synaptic unreliability on the information transmitted by spiking neurons. J Neurophysiol 79:1219–1229

Zhu JJ, Connors, BW 1999 Intrinsic firing patterns and whisker-evoked synaptic responses of neurons in the rat barrel cortex. J Neurophysiol 81:1171–1183

DISCUSSION

Logothetis: You have shown an impressive change of the membrane potential in the movement of the whiskers. Is this in contradiction with the imaging data that show that there is activation of one barrel which, after a certain period, spreads to any other barrels?

Brecht: These are two different things.

Logothetis: If the results I mentioned are to be taken as they were reported, they would imply that if you measure subthreshold activity the way you do with your patch clamping, a single cell should show a lot of unpredictable variations that are due to other whiskers. You are labelling one whisker, but this doesn't mean that the rat is only moving one whisker.

Brecht: The rat is moving all of them synchronously. At least when rats are whisking in air, this is the case. It is important to realize that the imaging data mostly come from anaesthetized animals. In awake animals, the population activity is more fractionated.

Logothetis: I'm not sure this is the answer. Whether the animals are anaesthetized or not, we are talking about connectivity and synergistic activity. The spikes can still be selective, but if the subthreshold for activity is some kind of weighted average, this is going to make the neuron you patch show much more unpredictable voltage changes than the single whisker you have.

Sparks: The 10 action potentials occurred over a short period, but the movement persisted for many seconds. What is happening here?

Brecht: When we stimulate extracellularly we see a brief twitch. It is very brief, even if we continue to stimulate. In awake animals this might be different. Secondly, in 20% of cases of extracellular stimulation we observe single whisker movements. However, we never saw this with intracellular stimulation. There is not a single case where we saw this brief twitch. What I think is happening is that cortical microstimulation and intracellular stimulation are activating the cortex in quite a different manner. My guess would be inhibition. The single whisker movement we observe with extracellular stimulation might be a result of inhibition. We also know that after we terminate the microstimulation there is enormous inhibition. I would think that perhaps this terminates the movement.

Derdikman: I was wondering whether there is a connection between the result you got in awake rats and what I was presenting for the anaesthetized rats. The movement-modulated cells are like whisking-touch cells in the work of Szwed et al (2003), while movement suppression cells are more like touch neurons. They do not respond to the whisking, while if there was an object in the path of their principle whisker, they may respond.

Brecht: We can't tell. I was favouring the idea that the excitation is restricted to a few cells that go with it and that have different phases.

Tanifuji: Physically, the whiskers are connected each other. If you stimulate a single whisker, isn't it the case that nearby whiskers are also stimulated?

Brecht: The follicle is designed to isolate the whiskers mechanically. At the same time rats can move whiskers individually.

Tanifuiji: There are upstate and downstate in membrane potentials, and spontaneous movement is higher at the downstate than at the upstate. Are these state-dependent changes in spontaneous movement synchronized across the whiskers?

Brecht: Yes. I think you could even record the synchronized up and downstates even in visual and in barrel cortex. The field potential looks like a mirror image of the subthreshold intracellular potential. This must mean that it is all the neurons joining an upstate. Up and down states are very much a population phenomenon, at least in motor cortex.

Logothetis: I was surprised to hear that you say that the field potential mimics the intracellular recordings. I would expect that no matter where you land with your electrode, capturing also the after-potentials and membrane oscillations would make the general form that you measure different from what you measure intracellularly.

Brecht: Perhaps this was an overstatement. But the striking thing is how similar intracellular recording, field potential and even the EEG are. There are blips in the membrane potential and blips in the EEG. They are not always the same size: there might be a big blip in the EEG and it is smaller in the cell. There is something surprising about the similarity of this ongoing activity, at least. I am not sure that this will hold for evoked activity, because neighbouring cells can differ quite a bit in their sensory properties. This statement would have to be restricted to ongoing activity. On the one hand we feel that every cell we record from is a bit different, and then we look at the field potential and the intracellular reading and they are very similar. Subtle differences in the single cell and the population activity might be important.

Logothetis: Is the barrel cortex characterized by the same very strong recurrency that we see in the primate cortex? There are a lot of collaterals of pyramidal cells that project onto either the excitatory interneurons or themselves. Is this scheme that predicts a very high gain response common in rats in the barrel cortex?

Brecht: I would say it is similar, but the primate cortex has not been studied as rigorously on the synaptic level. We know much more about *in vivo* recordings but there is much less slice work. In the slice work we have a very good feel for the connections between the neurons in the barrel cortex. We know how big the layer 4/layer 4 connection is and how big the layer 4/layer 2/3 connection is. One thing that looks a bit different is that a prominent feature of visual cortex, if you inject, is patchy connections. If you inject in barrel cortex it is also a bit patchy, but not as much as visual cortex.

Schall: In the microstimulation experiment, what was the latency to evoke a movement?

Brecht: It varies tremendously with the number of action potentials. With 10 it was reasonably fast, often in the first or second frame. Going to fewer action potentials it gets very long. With one or two, it takes many seconds. The general way we think about our effects is that in the brainstem we kick off a central pattern generator which has slow time constants to start and end with.

Reference

Szwed M, Bagdasarian K, Ahissar E 2003 Encoding of vibrissal active touch. Neuron 40:621–630

General discussion I

Treves: Mathew Diamond said at some point that he believes that the primary somatosensory cortex has all the information that rats need for discrimination. Perhaps one could say the same for the thalamic nuclei.

Diamond: Yes, I meant it that way.

Treves: The cortex may have some more information about movements and the state of the animal, but it is hard to think of what the cortex adds to earlier stages of processing without taking memory into account. I wonder to what extent analysing the encoding of stimuli—the representation of currently perceived stimuli—is able to shed light on cortical function if memory is not taken into account.

Diamond: Just to clarify my answer to the question of whether all the information is there, I meant it in the most simple-minded way. What the rat needs in order to make the decision must be there.

Wolpert: That isn't necessarily true, is it? If it needs the efferent copy to interpret the signal coming in, it doesn't have this information in the primary somatosensory cortex.

Diamond: I am making the assumption it has the same afferent copy whether it is touching one texture or the other.

Wolpert: If it needs to know how it is moving its body then this might not be in the primary somatosensory cortex either.

Diamond: Sensory coding is a necessary step towards understanding decision-making, but we also have to think about memory. I have a hypothesis that I am anxious to test in the trained animals: the rat has been trained on this task for many weeks and it probably doesn't have many other interesting events in its life besides this. It has one exciting hour in the maze where it is in an interesting environment and is getting food and so on. Every time it is put into the maze it is discriminating between the same two textures. My guess is that once it is trained it is not operating purely on afferent signals, but rather is forming expectations of what it is touching. Rather than purely waiting for sensory input, the rat may form a mental image and compare ongoing inputs to this. The decision is therefore not based purely on the input but on the comparison. I am interested in seeing whether, as the rat prepares to make the contact, there is some neuronal representation that is a pre-image, or expectation of what it is about to touch. If, then I would agree that in doing the task the rat is using its memory together with the afferent signal.

Wolpert: I have heard people like Ken Johnson say that there is no advantage in active sensing of tactile sensations over passive. In the rats, is there any evidence that if you immobilize their whiskers somehow, that there is any detriment to their performance? If they can't whisk, do they have any problems?

Brecht: I think there are a few negative results. There is very little behavioural work. Is active sensing an advantage?

Wolpert: In humans it is claimed that you are no better at sensing surfaces if you feel them actively than when the surface is passively moved under your finger.

Diamond: That statement is strongly debated.

Derdikman: Perhaps the comparison with whisking should not be made with our tactile system but instead with our eye movements. No one would argue that if our eye movements are paralysed we will see the same as when they are not. I would guess the same is true for whisking in the rat. The active movements are important.

Wolpert: So is this an accepted truth for the whiskers, and people don't think it is worth testing?

Scott: If you passively lock the whiskers, the rat could just move its head.

Wolpert: But the whiskers move with high frequency: I'm sure the rats can't move their heads at 10 Hz.

Romo: I wonder whether Dori Derdikman has observed neural signals in the primary somatosensory cortex which reflect the comparison process; in other words, a combination of top-down signals and bottom-up signals. If the primary somatosensory cortex contains all the neural computation for sensory discrimination, then the sensory input and some top-down signals such working memory, attention signals must be combined in the primary somatosensory cortex.

Derdikman: Just to be sure that I understand you, you mean a touch signal versus the whisking signal, where there are cells in which these are combined?

Romo: Yes.

Derdikman: Basically, in the cortex we have preliminary results that demonstrate that the majority of cells respond both to whisking while demonstrating an additional response related to touch. This means that many cells are of the Whisking-Touch type (Szwed et al 2003).

Romo: There is information encoded in the whisking and from touch, but is there some signal that matches both?

Derdikman: The point is that if there is a touch component then you get an additional spiking. The peristimulus time histogram (PSTH) demonstrates a higher amplitude when a touch component is present. It is as if in most of these units the touch component is riding on top of the basic whisking component

Sparks: I'd like to pursue the eye movement analogy. Yes, if you immobilize the eye the percept disappears. But there is an adverse consequence of moving the eye: if we look at contrast sensitivity function, when the image moves across the retina even at slow rates some high spatial frequency is lost. The way I am listening to the whisking story is that you would expect to gain high spatial frequency with whisking, not lose it. If this is true, then the analogy has limitations.

Wolpert: Having a fovea is a clear difference: the eye has a sensitive area whereas with whisking this isn't present.

Derdikman: There are cases where the rat is focusing its whiskers to a smaller field: in a sense, this is its fovea. The rat can also whisk at a higher frequency and slower amplitude, which is termed 'foveal whisking' (Berg & Kleinfeld 2003).

Haggard: Are all whiskers equally innervated, or are some particularly important, with a better representation? There could be two ways in which whiskers resemble the fovea. There is the active one where sensory resources are put into the most useful place by moving whiskers, but there is also a purely passive one with the best receptors located in certain areas.

Derdikman: The barrels are not similar in size.

Brecht: The longer the better for whiskers: the longer ones have more innervation.

Haggard: Isn't this a problem for the supposed important role of expectation? I would have thought that the front whiskers were the ones telling you about what is coming your way.

Brecht: If there are too few fibres then no barrel is formed. The more anterior whiskers indeed have smaller barrels, but they are more numerous. If rats are interested in an object they will invariably touch it with these small ones, too. For roughness the long ones are important, but in other experiments where the rats have to differentiate triangles from squares and so on, they can't do this with the long whiskers. If we cut the long whiskers they will happily eat the squares, but if the short ones are shaved they don't recognize the squares.

Logothetis: Mathew Diamond, were you using a real discrimination task or always a discrimination of smoothness of texture?

Diamond: In the behavioural experiments, the rats had to distinguish between smooth and rough surfaces in order to find the baited arms of a maze.

Logothetis: So it is like a detection of contrast in the visual system. It would be interesting to see whether the rat can discriminate different textures.

Diamond: That will be a future experiment.

Logothetis: I don't have enough background to do this myself, but is there an emerging principle here regarding the decision making from the first three papers we have heard?

Romo: I wonder whether columnar processing has been addressed during texture discrimination, in animals free from anaesthesia.

Brecht: There is very good evidence from Diamond's work that the topography is absolutely critical. An emergent theme is that most people in the barrel field are hesitant to call cortical variability 'noise'. My sense of the visual field is that this is often done. A grating is presented 100 times and results in 100 different responses, so they say it is very noisy.

Logothetis: It depends what you mean. If you mean that the cortical variability is to a large extent affected by a lot of subcortical gating signals from the brain stem or thalamus, everyone would agree with that. If you mean something beyond this, you need to clarify it. Where is the surprise? Cortical neurons show variability in their activity, and part of this variability can be explained by other signals that we don't investigate. Perhaps an additional variability is that there is indeed noise.

Brecht: This is a crucial question. What is the origin of cortical neuron variability?

Logothetis: You showed some things that we could usefully discuss. For example, you showed that spikes are relatively rare, for example. On the other hand, if you take into account the results from people not represented here, you would think that the columns are columns in some sense, but also they are very interactive with other columns. It seems that all these interactions lead in the end to the production of very few spikes. Are these few spikes, wherever they go, able to generate any movement?

Sparks: We have heard about a pattern generator. Where is it? In the cortex?

Diamond: It is in the brainstem.

Sparks: The 10 spikes must be interacting with the pattern generator to produce the prolonged movement. What do we know about the connections and interactions between the pattern generator and the cortex?

Brecht: Not enough. There are a few candidate circuits. It was thought to be the case that there was no output to the motor neurons, but this does occur. There is a sparse direct connection from motor cortex to the facial nucleus. Most of the cortical output does not go there, though. There are good candidate circuits for the central pattern generator for whisking. There is also good evidence that serotoninergic cells are involved in that circuit.

Diamond: About the general question of decision making, I don't think we have too much to say about this now. Michael has shown some important principles, though. For example, we now know that spikes might matter a lot, that individual spikes might be significant. But I think the main message is that we are just developing the methods to work out decision making in this sensory system. It is a system that hasn't been well studied: people haven't really looked at rats exploring their environments and making decisions. We are beginning to develop the tools that will give us some results in the next few years.

Schall: You published a paper showing that the signal from something like 10 cells in the barrel cortex of the rat was enough for the rat to know what it was going to do (Petersen et al 2001). Isn't this the case?

Diamond: We used Mutual Information to show that 10 cells carry enough information to give the answer about stimulus location—which whisker was contacted (Petersen et al 2001).

Schall: Observations made in area MT (Shadlen et al 1996) and in the frontal eye field (Bichot & Schall 2002) indicate that signals from as few as 10–100 neurons can signal stimulus properties. It seems that this occurs because those 10–100 neurons are representative of the population among which a sufficient degree of correlation ensures that a sample from any member of the population provides a good measure of the state of the entire population. Does this reconcile with the sparse coding conclusion? They seem to be in opposition.

Diamond: No, we sample cells in very different ways. Michael Brecht is using patch clamp, so he samples many cells that are not responsive, whereas we are using extracellular recording so cells carry information in our sample. If they don't give spikes we don't even notice them. Even so, it may be that we see the 10 responsive cells out of a few hundred cells. We have done some notes on papers comparing his sparseness with our observations, and they do match up well if we make reasonable assumptions about the proportions of cells actually involved.

Rizzolatti: Can you explain better how the system is organized? Did I understand correctly that most fibres reach an oscillatory circuit? If this is so, is it really surprising that there is no immediate response?

Brecht: Cortical layer 5, which is expanded in vibrissae motor cortex projects to brain stem centres. It is a set of nuclei, part of the raphe and various nuclei directly adjacent to the facial nucleus.

Rizzolatti: Can you obtain movements of a single whisker?

Brecht: I have a suspicion that there is very sparse input from motor cortex directly to the facial nucleus. In the primates there is the idea that single finger movements are well correlated with direct cortico–motor neuron projections. In other mammalian motor circuits we don't find this.

Rizzolatti: In most mammals the corticospinal tract does not terminate in the spinal chord lamina IX, where motor neurons are located, but in the so-called 'intermediate zone'. The cortical control of alpha motor neurons is done via interneurons.

Brecht: Interestingly, the direct input to facial nucleus is very focal. The motor cortex makes diffuse input to a set of nuclei in the brainstem. A small injection into motor cortex gives diffuse labelling in several brainstem nuclei. It is a good assumption that these form the rhythm generator. The input that goes directly to the facial nucleus is different. It goes to a very small spot and it is always the best whisker. If we back label the follicle we find that it goes to the best whisker of that motor cortical site. This is why I thought that the connection for single whiskers is similar for the monkey or human equivalent. We have no proof for this yet.

Logothetis: What are the prominent cortical inputs to the somatosensory cortex?

Brecht: M1 and S2 are big inputs, whereas M2 is very small. S1 is very heterogeneous. There are the barrels which are very different from the septa in between.

Logothetis: The reason I am asking is because of what Mathew Diamond said about all the information being there: it needs to be qualified whether there are feedback connections or not. Anything I would understand in the strict sense of information would pertain to feed forward channels. The moment you start having loops, it is very hard to say what the primary somatosensory cortex is having, because it is having a lot of everything.

Porro: A related question is where does the efference copy mechanism come from in this system? Is it from the primary motor cortex, M1? Or is it an input from the brainstem generator? How precise is it?

Brecht: I don't know.

Rizzolatti: What about SII? From SII there are several descending projections. Is it known where they go?

Brecht: I don't know.

Romo: Unless they go to the dorsal horn nucleus, to control the entry of information.

Diamond: There is also a poorly defined area behind barrel cortex, between the barrel cortex and visual cortex, which may have some analogy or homology to posterior parietal cortex. No one knows anything about this. There are projections to and from there both from visual cortex and whisker cortex.

Logothetis: Do you see the usual pattern of layer 5,6 neurons projecting to columns and other subcortical structures? Is there anything like driving cells and modulating cells in the rat?

Brecht: Layer 6 goes to the thalamus. There are a few specializations: layer 5 goes to some special thalamic nuclei, for example. But by and large the connection pattern is very similar to the visual system.

Diamond: Layer 5 also projects to striatum, cerebellum and superior colliculus.

Treves: You mentioned in passing the interesting discrimination tasks done by the myastacial vibrissae and the front vibrissae.

Brecht: Some tasks. They have to discriminate a small triangle from a square. They invariably do this with their small whiskers.

Treves: You also said that their organization in terms of barrels is different.

Brecht: The barrels are smaller and they are very well delineated. Layer 4 is bit different in big and small barrels, with lots of stellate cells in small barrels whereas in the big barrels there are a lot of pyramids.

Diamond: I wouldn't make the distinction you do between what the rats do with the small whiskers in the front and the larger ones at the back. The ones at the front are very densely packed. In the task Brecht used, where they had to discriminate a bitter cookie from a sweet cookie, and the two cookies had different shapes, they could detect the shape because the short whiskers are so densely packed that how they push against the whisker gives an imprint of the shape. The large whiskers pass over the surface in a disorganized way. It is hard to see how whiskers distant from each other could tell the rat anything about the shape of an object that would fall between the whiskers. So the large whiskers are better at things like texture discrimination and distance assessment.

Brecht: They are also larger in the brain: two thirds of the barrel cortex is taken by the large whiskers.

Haggard: That sounds a bit to me like a 'what' and 'where' distinction. The big whiskers are very much about location in egocentric space and distance from walls, and the front whiskers are concerned with object forms.

Diamond: They are concerned with forms of small objects. The large whiskers are good for forms of large objects, if we consider the forms of large objects to be things like the curvature of walls.

Haggard: Maybe the what/where distinction is a matter of scale, then.

Barash: You mentioned in passing the subject of coding biases. Could you go into more detail about this comparison you did between the results? What does 'coding bias' mean in extracellular and intracellular contexts?

Diamond: In extracellular recording there is a bias to studying neurons that respond to the stimuli. We don't stop the electrode to record from a cell if we don't know the cell is there.

Barash: You said that you did specific comparisons that led you to specific conclusions. Could you elaborate?

Diamond: The main thing we wanted to see was whether Michael Brecht's idea about the sparseness of coding is compatible with the spike counts we found with extracellular recording. If we were recording from the cells that were for him the ones that fired more frequently, this could account for our data. There was no incompatibility. In other words, we weren't drawing different conclusions about neuronal responses: we were just studying different populations of neurons. We found, on average, a larger value of spikes/stimulus than did Michael Brecht. Then when we excluded from his dataset the neurons that didn't fire any spikes per stimulus—neurons we would not even notice—then the spikes per stimulus were the same.

Brecht: I think we go wrong if we assume that whatever we record with our methods is necessarily representative. There will always be sampling biases. As long as we don't make this assumption, we'll be fine. But if we make calculations assuming sampling is representative, we'll end up with absurd assumptions about what is happening in the cortex.

Derdikman: Related to what we asked at the beginning about memory, I wanted to point out that when there were multiple whisks there was some memory, and preliminary results indicate that the cortical response at steady state was different from the response at the beginning. It was as if the response already knew that there will not be an object there and as a result there was an earlier latency. Thus it seems that also in the case of an anaesthetized rat, where mechanisms are not functioning fully, there seems to be some kind of simple memory in the cortex. Second, and more generally, there is the question of variability. It seems as if there is some inverse correlation between the receptive field size and variability. If you go to the trigeminal ganglion where the receptive fields are very small, you get very nice reproduction of your responses. If you go to the cortex where receptive fields are large, you get large variability.

Logothetis: This could be just correlation. Both things correlate with something else.

Brecht: In your hands, would an inferior temporal cortex (IT) cell be any more variable than a V1 cell, trial by trial?

Logothetis: I have no idea what an IT cell is doing. I don't think anyone is making big stories out of this because we know that we have a very biased narrow-angle view of the activity of these cells. The neuron integrates information from 10–15 000 synapses, and not all of these synapses will be related to the previous stage of processing, so you would expect to see variability.

Esteky: But the variability of cell responses in V1 is reported to be similar to MT (Shadlen & Newsome 1998) and IT (Gershon et al 1998) so the variance does not seem to increase along the cortical feed-forward processing.

References

Bichot NP, Schall JD 2002 Priming in macaque frontal cortex during popout visual search: feature-based facilitation and location-based inhibition of return. J Neurosci 22:4675–4685

Berg RW, Kleinfeld D 2003 Rhythmic whisking by rat: retraction as well as protraction of the vibrissae is under active muscular control. J Neurophysiol 89:104–117

Gershon ED, Wiener MC, Latham PE, Richmond BJ 1998 Coding strategies in monkey V1 and inferior temporal cortices. J Neurophysiol 79:1135–1144

Petersen RS, Panzeri S, Diamond ME 2001 Population coding of stimulus location in rat somatosensory cortex. Neuron 32:503–514

Szwed M, Bagdasarian K, Ahissar E 2003 Encoding of vibrissal active touch. Neuron 40:621–630

Shadlen MN, Newsome WT 1998 The variable discharge of cortical neurons: implications for connectivity, computation, and information coding. J Neurosci 18:3870–3896

Shadlen MN, Britten KH, Newsome WT, Movshon JA 1996 A computational analysis of the relationship between neuronal and behavioral responses to visual motion. J Neurosci 16:1486–1510

Switching of sensorimotor transformations: antisaccades and parietal cortex

Shabtai Barash and Mingsha Zhang

Department of Neurobiology, Weizmann Institute of Science, Rehovot 76100, Israel

Abstract. The sensorimotor processing necessary in complex realistic situations goes beyond straight-forward application of a given sensorimotor transformation. Contextual information may make it necessary to switch to another sensorimotor transformation. We studied the issue of switching by using the mixed memory prosaccade/antisaccade task. Neurons of the lateral intra-parietal area (LIP) might be involved in computing the sensorimotor transformations for both prosaccades and antisaccades. LIP neurons may also be involved in switching to the antisaccade sensorimotor transformation, when an antisaccade is requested. Some neurons in LIP show a paradoxical pattern of activity—motor in space but visual in time. Funahashi, Chafee and Goldman-Rakic reported in 1993 a complimentary pattern of activity in prefrontal cortex—visual in space but motor in time. These odd observations are explained by the hypothesis that (1) the parietal cortex contains a sensorimotor transformation module, and prefrontal cortex a context categorization module, and (2) following target onset, information flows from early visual system to parietal cortex and on to prefrontal cortex; then, a second wave of activation, contingent on a switching signal, arrives back at parietal cortex. The duration of this loop is less than 100 ms. Thus, the paradoxical activities are intermediate representations derived in the cognitive processing involved in switching sensorimotor transformations.

2005 Percept, decision, action: bridging the gaps. Wiley, Chichester (Novartis Foundation Symposium 270) p 59–74

Sensorimotor transformations

Much of this Novartis Foundation Symposium on 'Percept, decision, action: bridging the gaps' is dedicated to processing within the 'sensorimotor loop', illustrated by the case of vision and eye movements by the black arrows in Figure 1. Images of the world falling on our retinas evoke saccadic eye movements. Saccades evoked by the previous retinal image determine the subsequent retinal image—most importantly, by determining what spot in the image would fall on the fovea and hence be processed with much higher acuity than the rest of the visual scene. Thus, visual percepts build up within the framework of a sensorimotor loop made up of

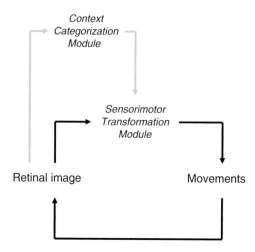

FIG. 1. The sensorimotor loop (black arrows) must be complemented by the context categorization loop (grey arrows).

retinal sensations and saccadic eye movements (and, sometimes, additional components).

Sensorimotor transformations are closely related to sensorimotor loops. Sensorimotor transformations implicitly refer to space. Consider the input and output of the brain's processing of visually guided movements: the input is a series of activations of retinal neurons, the output a finely tuned and accurately timed activation of the fibres of the extraocular muscles. Although very different from each other, both input and output share one fundamental attribute, which relates to *space*. In the case of a simple saccade, aimed to shift the fovea toward the image of a small, well defined, isolated object ('single target'), one can think of a 'visual vector' starting at the point currently fixated by the fovea and ending at the isolated object (Fig. 2a, top row). The movement defines a motor vector, starting at the pre-movement eye position and ending at the post-movement eye position. The objective of the sensorimotor processing is to generate a movement whose motor vector would be equal to the visual vector. Thus, a foveating saccade is characterized by a sensorimotor transformation of *identity*.

Are sensorimotor transformations computed directly, in one step, or in several steps—inevitably involving intermediate representations between visual and motor? In 1975, Mountcastle and collaborators proposed the 'command neuron hypothesis': 'the posterior parietal cortex regions contain a command apparatus for operation of the limbs, hands, and eyes within immediate extrapersonal space. This general command function is exercised in a holistic fashion. It relates to acts aimed at certain behavioural goals and not to the details of muscular contraction during

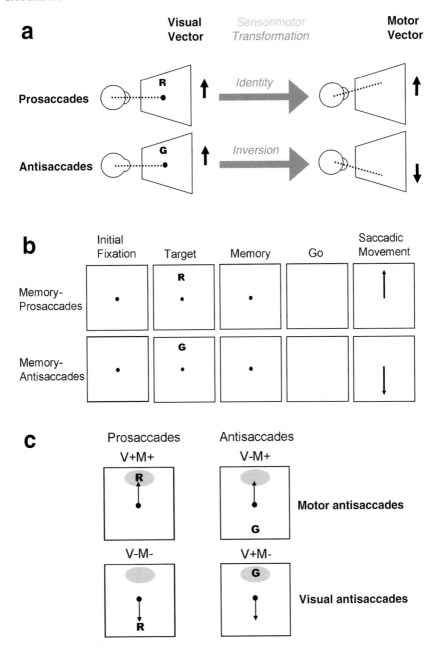

FIG. 2. The mixed memory prosaccade/antisaccade task. (a) schematic of the involved senso-rimotor transformation. (b) Stages of the tasks. (c) Four types of trials are defined with respect to a given neuron's response field. Figs 2b&c reprinted with permission from The American Physiological Society, Zhang & Barash (2004).

execution. These details are, on this hypothesis, made precise by the motor system, for which it is well suited by virtue of its powerful mechanisms for specifying movement exactly' (Mountcastle et al 1975). Thus, Mountcastle and collaborators suggested a specific sort of intermediate representation. We now know that many cortical areas and subcortical regions appear to contain holistic motor-intention signals related to saccades (holistic in the sense that they are coded spatially, not in muscle coordinates). Are these holistic signals themselves computed in one step, or are they built up gradually, implicating the existence of intermediate representations? To approach this question, one must focus on the cognitive processing in which these holistic intention signals are involved. We argue that, with this perspective, intermediate representations can indeed be found. One intermediate representation is the 'paradoxical activity' we described in area LIP (Zhang & Barash 2000, 2004). Another type of intermediate representation was described in prefrontal cortex (Funahashi et al 1993). This odd prefrontal pattern of activity complements in a surprising way the paradoxical activity we described in the parietal cortex.

The switching problem

Visual–motor processing and visual–motor transformations become *much* more complicated when the visual scene contains more than a single, small object. Imagine a forced takeover of a room occupied by kidnappers intermingled with hostages. A main objective is to aim toward the kidnappers, but, at the same time, away from the hostages. Even a single object may require a rapid response that varies drastically depending on contextual knowledge. Imagine upon entering a room that a person is detected in the room. Kidnapper or hostage? Friend or foe? This is a life-and-death decision that must be resolved instantly and followed by opposing actions: if a kidnapper, attack; if a hostage, bring to safety. What is the decision based on? The cues might be subtle or simple, such as the presence of a gun, even cloth colour. The pattern of interpretation and response might easily be more complicated: imagine, upon entering a room, observing a single person that appears to be a hostage and is covertly pointing at a hidden corner of the room.

Complex situations similar to the kidnapper–hostage scene (even if less dramatic) occur regularly in social interactions of both humans and monkeys. Thinking about these situations leads us to suggest that the sensorimotor loop of Figure 1 must be appended. In parallel to the 'main route', by which the current sensory input feeds into the active sensorimotor transformation, there must be an additional route that inspects the image from a more general perspective, integrating additional, contextual information. The contextual information may necessitate making an entirely different motor response to the very same stimulus. Thus, the previously active sensorimotor transformation might have to be switched off, and an alternative sensorimotor transformation switched on instead. Thus, the contextual information would instigate switching of sensorimotor transformations.

These considerations lead to the hypothesis illustrated in Figure 1 (Barash 2003). Two 'modules' are postulated to be present. A 'sensorimotor transformation module' implements the active sensorimotor transformation, and is also capable of switching to other sensorimotor transformations stored in memory. But what makes the sensorimotor transformation module switch from one transformation to another? There must be some signal commanding the switch. The command signal depends on contextual information. We are thus led to postulate a second, 'context categorization module'. Based on the visual input (as well as other sources of information), context is categorized, an appropriate command signal is generated, and is sent to the sensorimotor transformation module, instructing a switch to the appropriate sensorimotor transformation. These hypothetical modules and the flow of information between them are illustrated in the grey arrows of Figure 1 (Barash 2003).

The switching problem relates to these hypothetical modules. What is the nature of the context-contingent categorization? How are sensorimotor transformations actually switched? Where in the brain do these modules exist? Can switching-related signals be detected at all?

The mixed memory prosaccade/antisaccade task

One experimental paradigm that captures the core of switching in laboratory conditions will be described now. First, some terms: saccades can be made to the remembered locations of targets presented earlier in the trial ('memory saccades'), or made voluntarily, as, for example, in 'antisaccades', that are eye movements directed opposite the target (Everling & Fischer 1998, Hallett 1978). In the context of antisaccades, foveating saccades are sometimes called 'prosaccades'.

As discussed above, the sensorimotor transformation of foveating saccades is that of identity (Fig. 3a, top row). Antisaccades are marked by a sensorimotor transformation of *inversion*: the objective of the sensorimotor processing is to generate a movement whose motor vector is the inverse of the given visual vector. Schlag-Rey and collaborators, who first described this transformation, asked where in the brain, and how, this transformation is computed. They argued that it is not in supplementary and frontal eye fields (Schlag-Rey et al 1997). Below we suggest that the transformation might be computed in the LIP (Zhang & Barash 2000, 2004).

Figure 2b is a scheme of the variant of the mixed prosaccades/antisaccades task that we have used (Zhang & Barash 2000, 2004). All trials are of memory saccades, involving working memory. The use of memory is experimentally beneficial: a sufficiently long memory prevents visual and motor processing from overlapping in time. The target's colour is the contextual cue that instructs the monkey to make

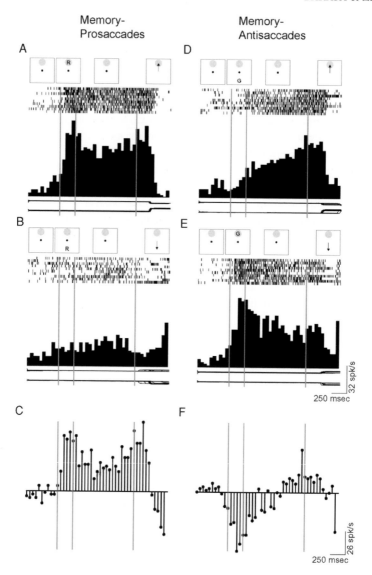

FIG. 3.   Activity of an LIP neuron in the memory prosaccade/antisaccade task. (A–C) Memory prosaccades, (D–F) memory antisaccades. (A,B,D,E) Each panel illustrates, from the top, the spike raster and histogram, and superimposed eye position records from the same trials as the raster. The vertical dotted lines mark, from the left, the onset and offset of the target, followed by the offset of the fixation spot. This offset serves as the 'go signal'. (C, F) Differential activities, showing the activity in A minus B, and in D minus F, respectively. Reprinted with permission from The American Physiological Society, Zhang & Barash (2004).

either a prosaccade (red target) or an antisaccade (green target). LIP neurons are generally not colour sensitive; hence the difference in colour is not likely to affect results—certainly not on the population level. Except for the target's colour and the direction of the saccade, in all other ways the prosaccade and antisaccade trials are the same.

Every trial of the mixed memory prosaccade/antisaccade task requires the subject to compute a sensorimotor transformation. Many trials require the sensorimotor transformation to be switched to the antisaccade configuration. Does the neuronal activity reflect the computation and switching of the transformations?

To approach this question, we studied neurons with trials of four types (Fig. 2c). These types are defined with respect to a neuron's *response field* (schematically represented by the grey ellipses in Fig. 2c). In 'on-field' prosaccades (Fig. 2c, top left panel) the target falls in the response field and the saccade is directed toward the response field; in 'off-field' prosaccades (bottom left panel), both target and saccade are directed away from the response field. The difference between the response in on-field and off-field prosaccades reflects the response-field-specific activity of the neuron. Is the response-field-specific activity visual or motor? Antisaccades help resolve this issue, because they come in two variants: in 'motor antisaccades' (top right panel) the visual target falls outside the response field, and the movement is made toward the response field. Hence the choice of name: the observation of activity in this type of antisaccades suggests that the activity is motor indeed, not visual. Similarly, in 'visual antisaccades', the target but not the movement falls in the response field. Observation of activity in visual antisaccades suggests that that activity is visual, not motor.

Parietal cortex neurons that might compute the sensorimotor transformation for antisaccades

Area LIP has been associated with saccades, selective visual attention and associated cognitive processing for a long time (Andersen 1987, Andersen et al 1997, Bisley & Goldberg 2003, Bisley et al 2004, Bracewell et al 1996, Bushnell et al 1981, Colby et al 1996, Gnadt & Andersen 1988, Gottlieb & Goldberg 1999, Gottlieb et al 1998, Snyder et al 1997, Thier & Andersen 1998). Area LIP has in particular been suggested to play a role in sensorimotor transformations (Andersen 1987). Our data suggest that some LIP neurons might be involved in computing the sensorimotor transformation for antisaccades. As described in the previous section, the core of this transformation is vector inversion.

Figure 3 shows a neuron that might contribute to this computation. In memory on-field prosaccades this neuron is activated shortly after the onset of the target, and remains active throughout the memory interval until the saccade is made. The neuron is active throughout the trial in *both* motor (Fig. 3D) and visual (Fig. 3E) memory antisaccades. However, the time-course of the activity is very different in visual and motor memory antisaccades. In visual memory antisaccades the neuron is briskly activated by the visual stimulus; later on, the response slowly decreases. In contrast, in motor memory antisaccades the activity builds up slowly, throughout the memory interval. Thus, in memory antisaccade trials the activity of this neuron changes from predominantly reflecting the direction, initially, of the visual target, to, eventually, of the intended movement. This change is captured in Figure 3F, which shows the 'antisaccade differential activity': the difference, per time in the trial, of the activity in motor and visual memory antisaccades. In spite of differences in the responses of individual neurons, analysis of the mean activity of many neurons shows that the mean population response also changes direction from visual to motor (Zhang & Barash 2000, 2004).

This pattern of activity is consistent with the possibility that area LIP is involved in computing the inversion from visual to motor, that is, the sensorimotor transformation of antisaccades.

The 'paradoxical', second wave of activity in the parietal cortex in memory antisaccades

Figure 4 illustrates a pattern of activity that, we believe, is related to the switching of the sensorimotor transformation module from the prosaccade to the antisaccade configuration. In on-field memory-prosaccade trials, the neuron discharges briskly after the target's onset. The activity goes down after the target's offset; at the time of the movement, the neuron is inactive. Thus, the timing of this activity is that of a visual response.

If this neuron's activity is indeed visual, then, when tested in memory antisaccades, we expect the neuron to show: (1) a similar response in visual memory antisaccades, and (2) no specific response in motor memory antisaccades. Prediction (1) does hold for this neuron. In visual memory antisaccades (bottom right panel of Fig. 4a) the neuron is also activated briskly shortly after target onset, and the response goes down after target offset, long before the saccadic movement. Thus, the neuron's response in visual memory antisaccades is consistent with the proposal that this is a neuron with a visual response.

However, prediction (2) fails. In motor memory antisaccades (top right panel of Fig. 4a) the neuron is vigorously active as well. Because the stimulus falls outside the response field this activity is not a visual response, at least not in the straightforward sense (of a classic receptive field).

What might explain this odd activity? Could it be a motor response? No, the timing of the response is not consistent with this activity being motor. Compare the response in motor memory antisaccades of the two neurons illustrated in the top right panels of Figure 3 and of Figure 4. The neuron of Figure 3 is only weakly

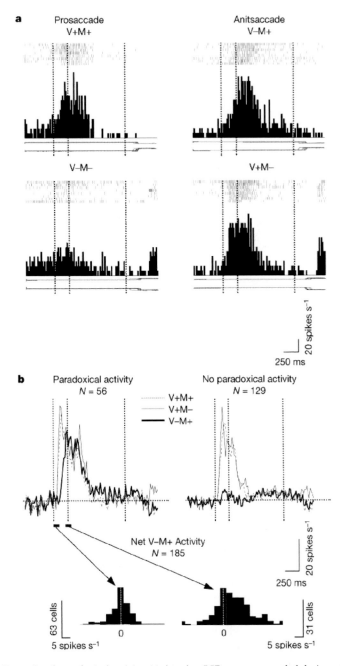

FIG. 4.   Example of paradoxical activity. (a) Another LIP neuron recorded during performance
of the same task. Same format as panels A,B,D,E of Fig. 3. (b) Mean activity in two groups of
visual neurons shows that about one third of these neurons show paradoxical activity. Reprinted
with permission from Zhang & Barash (2000).

active while the target is present. Only subsequently, throughout the memory period, the rate of spikes gradually increases, peaking close to the time of the saccade. This is the time-course of a typical motor-intention response, building up toward the intended movement. Now look at the response of the neuron of Figure 4. The response follows target onset with a brisk discharge (albeit with a slightly longer latency than this neuron's visual responses). After the target's offset the response goes down. By the time of the saccade the response is virtually back to baseline. This time-course is *not* consistent with the reflection of a motor intention. It is the time-course of a visual response.

Because the neuron's activity in *motor* memory antisaccades has a time-course that appears *visual*, and is certainly not motor, we call this response 'paradoxical'.

The amplitude of the paradoxical activity varies between neurons with visual activity. Nevertheless, the analysis presented in Figure 4b shows that, to a first approximation, about one third of the visual neurons do show paradoxical activity in motor memory antisaccades, in addition to their classic visual discharges.

Figure 4b further shows that the time-course of the mean paradoxical activity is similar to the time-course of the mean visual response, with one exception: the latency of the mean paradoxical activity is longer than the latency of the mean visual response by about 50 ms. Paradoxical activity is probably present also in neurons that, in addition to a visual response, have motor build-up activity (Zhang & Barash 2004).

What could the explanation for the paradoxical activity be? We suggest that the paradoxical activity reflects, directly or indirectly, the switching command that arrives in LIP from the hypothesized context categorization module. Because the paradoxical activity is present only in antisaccade trials, because the colour of the target is the only signal specifying whether an antisaccade is requested, the context categorization module must get the target's colour as input and generate (if the colour is green) a switching command signal that must quickly be fed back to the sensorimotor transformation module in LIP. In spite of our ignorance regarding the context categorization module, it appears that the postulated command signal *must* exist. How does the paradoxical activity relate to this postulated signal? Does the paradoxical activity reflect the arrival of the switching signal itself? An alternative explanation might be that the paradoxical activity is a non-classic visual response to a non-classic receptive field gated by the switching signal. Indeed, Schlag-Rey et al (1997) suggested that the inversion might be implemented by non-classical receptive fields that are gated in antisaccade trials. The time-course of the mean paradoxical activity (Fig. 4b) is consistent with their hypothesis. Even if the paradoxical activity is a non-classic visual response, it still must be triggered by a switching command, which must be generated at a context-categorization module. Thus, directly or indirectly, the paradoxical activity is contingent on a switching signal.

Comparison of the parietal and prefrontal paradoxical activities

The paradoxical activity appears to be motor in space but visual in time. What about the opposite pattern, visual in space and motor in time? Are there any neurons with this pattern?

In area LIP we have not observed neurons with this pattern of activity. Yet, surprisingly, this pattern was observed and reported more than 10 years ago in prefrontal cortex by Goldman-Rakic and collaborators (Funahashi et al 1993). These authors studied the activity of dorsolateral prefrontal cortex in a variant of the mixed memory prosaccade, memory antisaccade similar to the one we used. They studied neurons that in memory prosaccades show persistent activity building up toward the saccade. In memory antisaccades many of these neurons had the following response pattern: in motor memory antisaccades the neurons did not respond at all (above baseline). In visual memory antisaccades these neurons showed build-up activity that peaked close to the time of the saccades. Hence, visual in space, motor in time—we call this activity 'prefrontal paradoxical', even though this is not the originally used term (Funahashi et al 1993).

How can the prefrontal paradoxical activity be explained? The original report put forward functional suggestions that are in no contradiction with the following thoughts, spelled out in more detail elsewhere (Barash 2003). We suggest that the prefrontal paradoxical activity shows that the prefrontal cortex logically precedes or contains the context categorization module. It is a motor discharge in visual coordinates. This motor discharge remains in visual coordinates precisely because this neuron logically precedes (and may be part of) the context-categorization module. Because it precedes context-categorization, the neuron has no access to the output of the categorization process, that is, to the switching signal.

We are thus led into the following suggestion regarding the information flow in the mixed prosaccade/antisaccade task, illustrated in Figure 5. There are two waves of activation in area LIP. The first wave is of visual responses; it affects neurons with visual activity directly evoked by the target. The activity proceeds in two main streams. One stream proceeds on to other areas and downstream, probably via the superior colliculus to preoculomotor centres. In parallel, a second stream proceeds, perhaps vial parietal cortex, to other cortical regions and onto the context categorization module. Then, the output of context categorization, the switching signals, arrives at area LIP and evokes the second wave of activation, of paradoxical activity. The wave of paradoxical activity involves visual neurons that were not activated in the first wave of that trial. The two sets of activated neurons are largely disjoint, because their response fields are opposite each other. The second wave of activity, directly or indirectly, is contingent and reflects the switching command. For more details on this hypothesis see (Barash 2003).

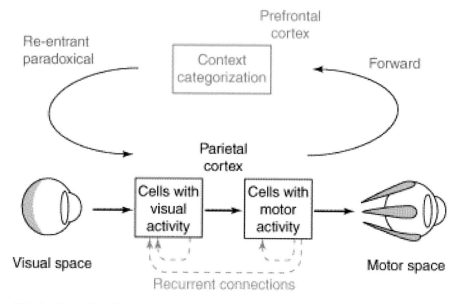

FIG. 5.   Illustration of the proposed flow of information implicated in the context-contingent switching of the sensorimotor transformation in the mixed memory prosaccade/antisaccade task. Reprinted from Barash (2003), with permission from Elsevier.

Conclusion

The mixed memory prosaccade/antisaccade task appears to be a good working model for studying switching of sensorimotor transformation by contextual signals. Area LIP contains neurons that might reflect the computation of the sensorimotor transformation in both memory-prosaccade and memory-antisaccade trials. Area LIP also contains visual and visual–motor neurons with a paradoxical pattern of activity—motor in space but visual in time. A complementary pattern of activity, visual in space but motor in time, was reported in prefrontal cortex. Both types of paradoxical activity can be explained in a straightforward manner by the hypothesis illustrated in Figure 5. The parietal cortex is presumed to contain the sensorimotor transformation module. Prefrontal cortex might contain the context categorization module.

References

Andersen RA 1987 The role of the inferior parietal lobule in spatial perception and visual-motor integration. In: Plum F, Mountcaste VB, Geiger ST (eds) Handbook of Physiology, Section 1: the nervous system. American Pysiological Society, Bethesda, MD, p 483–518

Andersen RA, Snyder LH, Bradley DC, Xing J 1997 Multimodal representation of space in the posterior parietal cortex and its use in planning movements. Annu Rev Neurosci 20:303–330

Barash S 2003 Paradoxical activities: insight into the relationship of parietal and prefrontal cortices. Trends Neurosci 26:582–589

Bisley JW, Goldberg ME 2003 Neuronal activity in the lateral intraparietal area and spatial attention. Science 299:81–86

Bisley JW, Krishna BS, Goldberg ME 2004 A rapid and precise on-response in posterior parietal cortex. J Neurosci 24:1833–1838

Bracewell RM, Mazzoni P, Barash S, Andersen RA 1996 Motor intention activity in the macaque's lateral intraparietal area. II. Changes of motor plan. J Neurophysiol 76:1457–1464

Bushnell MC, Goldberg ME, Robinson DL 1981 Behavioral enhancement of visual responses in monkey cerebral cortex. I. Modulation in posterior parietal cortex related to selective visual attention. J Neurophysiol 46:755–772

Colby CL, Duhamel JR, Goldberg ME 1996 Visual, presaccadic, and cognitive activation of single neurons in monkey lateral intraparietal area. J Neurophysiol 76:2841–2852

Everling S, Fischer B 1998 The antisaccade: a review of basic research and clinical studies. Neuropsychologia 36:885–899

Funahashi S, Chafee MV, Goldman-Rakic PS 1993 Prefrontal neuronal activity in rhesus monkeys performing a delayed anti-saccade task. Nature 365:753–756

Gnadt JW, Andersen RA 1988 Memory related motor planning activity in posterior parietal cortex of macaque. Exp Brain Res 70:216–220

Gottlieb J, Goldberg ME 1999 Activity of neurons in the lateral intraparietal area of the monkey during an antisaccade task. Nat Neurosci 2:906–912

Gottlieb JP, Kusunoki M, Goldberg ME 1998 The representation of visual salience in monkey parietal cortex. Nature 391:481–484

Hallett PE 1978 Primary and secondary saccades to goals defined by instructions. Vision Res 18:1279–1296

Mountcastle VB, Lynch JC, Georgopoulos A, Sakata H, and Acuna C 1975 Posterior parietal association cortex of the monkey: command functions for operations within extrapersonal space. J Neurophysiol 38:871–908

Schlag-Rey M, Amador N, Sanchez H, Schlag J 1997 Antisaccade performance predicted by neuronal activity in the supplementary eye field. Nature 390:398–401

Snyder LH, Batista AP, Andersen RA 1997 Coding of intention in the posterior parietal cortex. Nature 386:167–170

Thier P, Andersen RA 1998 Electrical microstimulation distinguishes distinct saccade-related areas in the posterior parietal cortex. J Neurophysiol 80:1713–1735

Zhang M, Barash S 2000 Neuronal switching of sensorimotor transformations for antisaccades. Nature 408:971–975

Zhang M, Barash S 2004 Persistent LIP activity in memory antisaccades: working memory for a sensorimotor transformation. J Neurophysiol 91:1424–1441

DISCUSSION

Schall: You have described these results in a particular way, but there is an alternative. For example, one can talk about the paradoxical activity in the frontal lobe as representing a rule. Steve Wise and Earl Miller have talked about prefrontal neurons encoding rules, and there is a rule being applied here. Thus, might the signal you observed not be movement activity, but instead just the rule being read out?

Barash: There is no contradiction here. Our interpretation is more specific. In categorizing context, prefrontal neurons by definition implement rules. On the other hand, the neurons we suggest do not compute just any rule. They specifically categorize context to lead (via the sensorimotor transformation module) to the appropriate response. We suggest a very specific hypothesis—representation of space prior to context-contingent transformations. (By the way, in computing sensorimotor transformations do parietal neurons not implement rules, too?)

Schall: Another alternative account of your observations in parietal cortex involves the allocation of attention. In your free choice task the two red spots were visually similar. Lots of evidence demonstrates that attention is allocated to stimuli according to their similarity to the target. So another account of your data may be that attention was allocated equivalently to both stimuli because the visual similarity prevents the activity from becoming different until later.

Barash: I agree with you. Attention is reflected in LIP as enhancement of the visual response; Goldberg and collaborators showed this ages ago. However, in the free-choice task we did not find any enhancement effects. It appears that attention is not a necessary aspect of choosing a target for a saccade.

Logothetis: I would like you to comment on two of your findings. First, what you call paradoxical activity is not identical to the visual response. I noticed that your visual activity is always substantially earlier. A decision is made as to whether the rule-based complex pathway is going to be used, or whether you are going to go directly to some kind of motor centres.

Barash: Right. The paradoxical activity has a somewhat longer latency than the visual response of the same cell. Yet its time-course is much more similar to that of a visual response than to a motor one. It comes on after stimulus onset and is down to baseline long before the movement. Does it reflect a decision? Yes, it must reflect a decision if it is observed in only part of the trials. The decision is exactly the task of the context-categorization module.

Logothetis: You called it 'paradoxical': I find it a little easier to explain. In the spot where you do the recordings, there will be cells like the ones that you reported, cells that are visual and motor. The visuals are going to give a signal that a sacchade will be initiated at some time. This signal will be available to other neurons that are not visual. They will appear to be timed. I don't see this as paradoxical. It is a sequence of connectivity between visual only, motor only, visual–motor cells and so on. Interestingly, whatever the sources of the information to get the cell going, after the cell starts there is a decision whether you should go through the frontal lobe or not. It seems that there is a time difference. Some of them start very early and others start almost 20 ms later.

Barash: Let me tell you in what sense I see these responses as paradoxical. By their time-courses these responses appear to be visual. Right, their latency is slightly longer, but this is LIP, not V1; visual latencies vary a lot from neuron to neuron.

Definitely, these responses do not reflect motor-intention. The activity of motor-intention neurons builds up toward the saccade; the paradoxical activity is gone long before the saccade. So, by their time-course, at least to a first approximation, they are visual.

On the other hand, nothing falls in their receptive field. Only the movement is made in their direction. So, in time they appear visual, in space motor. At least at first sight this appears an absurd, almost self-contradictory statement. In this sense it may be paradoxical—although the name 'paradoxical' in itself is of course not essential. We picked it up because this is how we felt when we first observed it.

Now, is it just a sequence of connectivity between visual and motor? I suspect not. You don't see it in prosaccades, which involve visual to motor connectivity as well. And why would you on the basis of a connectivity hypothesis expect parietal cortex to include cells with visual timing in the motor direction but not cells with motor timing in the visual direction? How would you explain the segregation between parietal and prefrontal cortex? I suspect you have to assume more than just visual to motor connectivity to explain these results.

Schall: Our lab has done a study of monkeys producing prosaccades and antisaccades in response to a visual search array (Sato & Schall 2003, Schall 2004). The shape of a colour singleton cues the monkeys which saccade to produce. Most of the visually responsive cells in frontal eye field select the colour target that is in the receptive field initially, but in antisaccade trials their representation changes to signal the endpoint of the antisaccade. On the other hand, about a third of the visual cells in frontal eye field only select the endpoint of the anti-saccade, so in the context of this visual search task this looks a lot like the paradoxical activity you describe.

Treves: It is not nice to ask about the work of the late Professor Goldman-Rakic, but your example was very clear cut, and theirs, at least as shown in your slides, was quite messy. This is not important for the discussion of your work, but it is important for your model with the role of the prefrontal cortex, where you said that the prefrontal cortex is before this context categorization stage.

Barash: This comment refers to Funahashi et al (1993). Prefrontal cortex is difficult to record and many reports show results that at first sight are not clear-cut. But I think their neuron is clear cut. It shows clear persistent, memory-interval activity in two of the four conditions, the on-field prosaccades and the visual antisaccades; no response in the other condition. The authors report that two-thirds of the cells have this pattern of activity. We are now trying to directly compare activity in the two regions.

Scott: I wanted to discuss your example where there were multiple targets and the monkeys chose one. I wanted to compare that result with a no-go task, where there is a target in the periphery that has been identified but you reward for not moving. LIP is one of the areas where you will often see activity to potential targets. Is there effectively the same response in that no-go task as is seen in yours?

Barash: Unfortunately we did not do a no-go task. There are recent reports in the literature. I agree it's an interesting experiment to do. However, there is a confound in no-go experiments that must be addressed. You tend to interpret activity as reflecting the decision not to go, but it may still be a motor plan that is simply not actuated.

References

Funahashi S, Chafee MV, Goldman-Rakic PS 1993 Prefrontal neuronal activity in rhesus monkeys performing a delayed anti-saccade task. Nature 365:753–756

Sato T, Schall JD 2003 Effects of stimulus-response compatibility on neural selection in frontal eye field. Neuron 38:637–648

Schall JD 2004 On the role of frontal eye field in guiding attention and saccades. Vision Res 44:1453–1467

Saccade initiation and the reliability of motor signals involved in the generation of saccadic eye movements

David L. Sparks* and Xintian Hu*†

*Department of Neuroscience, Baylor College of Medicine, One Baylor Plaza, Houston, TX 77030, USA, and †Laboratory of Sensory Motor Integration, Kunming Institute of Zoology, The Chinese Academy of Sciences, Kunming 650223, People's Republic of China

Abstract. We examined the trial-by-trial relationship between the metrics of saccadic eye movements and the activity of individual putative premotor neurons in the paramedian pontine reticular formation (PPRF) of rhesus monkeys. The region of the pons containing these excitatory burst neurons (EBNs) extends for several millimetres. Motoneurons innervating extraocular muscles integrate the output of hundreds or even thousands of these neurons. Accordingly, there no reason to expect that relatively small variations in the activity of a single pontine neuron would be related to variations in saccade amplitude or speed observed during repetitive eye movements to the same target. Nonetheless, we observed consistent relationships between variations in the number of spikes in the burst of pontine neurons and the amplitude of the saccade. Trial-to-trial variations in the instantaneous spike frequency during a burst are associated with variations in the velocity profile of the movement. Based on these data, we conclude that the activity of pontine burst neurons is not statistically independent and that simultaneous recordings from multiple cells will reveal a high degree of correlated activity.

2005 Percept, decision, action: bridging the gaps. Wiley, Chichester (Novartis Foundation Symposium 270) p 75–91

Saccade initiation

The neural mechanisms involved in the execution of a saccadic eye movement are fairly well understood (Sparks 2002). A saccade occurs when the motoneurons innervating the extraocular muscles produce a sudden, but brief, increase in discharge rate. This 'pulse' of motoneuron activity produces the transient increase in muscle tension needed to move the eye at saccadic velocity. The size and speed of saccades are controlled by the amplitude and duration of the pulse. A step change in the firing rate of motoneurons produces the steady state muscle tension needed to hold the eye at the new location in the orbit.

The signals producing the pulse and step of activation of motoneurons are constructed by feedback circuits found in the brainstem (Moschovakis et al 1996, Scudder et al 2002). Excitatory burst neurons (EBNs) located in the pons and rostral midbrain produce the pulse of activation. An instantaneous frequency plot of the high frequency burst is shown in Figure 1A (top left). The axons of EBNs project directly to the motoneurons innervating extraocular muscle fibres. The number of spikes in the burst of activity is highly correlated with the amplitude of the movement. Also, various temporal properties of the burst are correlated with temporal aspects of the saccade: burst onset is tightly coupled to saccade onset; burst duration is highly correlated with saccade duration; and the peak velocity of the saccade is related to the peak frequency of the burst (Fig. 1A, right).

The activity of EBNs is controlled, in turn, by omnipause neurons (OPNs). OPNs are clustered together (Buttner-Ennever et al 1988) in a small midline nucleus (nucleus raphe interpositus). They discharge at a relatively constant rate during fixation intervals but stop firing in association with saccades in all directions (Fig. 1A, bottom left). The cessation of activity (pause) begins before saccade onset, before the onset of the EBN burst, and ends before the saccade is terminated. OPNs produce monosynaptic inhibition of EBNs (Nakao et al 1980, Curthoys et al 1984, Strassman et al 1987) and saccades can be interrupted in mid-flight by microstimulation of the OPNs (Keller 1974, 1977, King & Fuchs 1977).

The question of saccade initiation now becomes: 'what is the "trigger" signal that turns off the OPN cells?'. The superior colliculus (SC) is a major source of inputs to the pulse-step generator circuits. Neurons in this midbrain structure form a map of saccadic eye movements that can be revealed using microstimulation methods. Electrical stimulation of the intermediate and deeper layers of SC of monkeys produces conjugate, contralateral saccades with amplitudes and directions that depend on the site of collicular stimulation (Robinson 1972). The site of stimulation within the map determines the largest movement that can be produced. However, for any stimulation site as the duration of the stimulation train increases, movement amplitude increases monotonically until it reaches the site-specific limit. Additionally, the peak velocity of the evoked movement is influenced by the frequency of stimulation; higher frequencies produce movements with higher velocities. The effects of train duration and frequency can be varied to produce movements that have comparable amplitudes but different velocity profiles. Similarly, stimulation parameters can be adjusted to evoke movements of the same amplitude with different latencies. These data (Stanford et al 1996) indicate that at least three independent signals are derived from the spatial and temporal pattern of collicular activity—one specifying the desired displacement, another related to saccadic velocity and a third involved in the initiation of a saccade.

One particular type of neuron found in the SC, the saccade-related burst neuron (SRBN), is thought to play a key role in saccade initiation. These neurons may

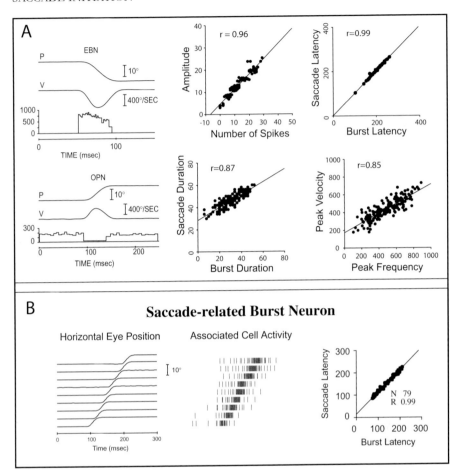

FIG. 1.   Role of pontine and collicular neurons in saccade initiation. (A) (*Left*) Instantaneous frequency plots of the saccade-related activity of an EBN (top) and OPN (bottom). P, horizontal position; V,: horizontal velocity. (*Right*) Plots of the relationships of measures of the activity of an individual EBN with saccade latency, amplitude, duration and velocity. (B) (*Left*) Plots of horizontal eye position for nine saccades ranked in order of saccade latency. (*Centre*) Raster plots of the activity of a SRBN during the nine movements. (*Right*) Relationship between burst latency and saccade latency.

display a low frequency prelude of activity before a saccade, but 18–20 ms before saccade onset, the low-frequency activity is replaced by a high-frequency burst of activity (Fig. 1C). In behavioural situations in which saccadic reaction time varies over large ranges, the onset of the burst of SRBNs is tightly coupled to saccade onset (Fig. 1C). Moreover, the occurrence or lack of occurrence of the high-frequency burst can be used to predict perfectly the behaviour of a monkey performing a behavioural task in which the probability of saccade initiation is manipulated by varying target duration (Sparks 1978). The axons of SRBNs form a major efferent pathway from the SC to subsequent oculomotor premotor neurons (Raybourn & Keller 1977, Moschovakis et al 1996). These and other findings form the basis of the hypothesis (Sparks 1978, Keller 1979, Sparks et al 2000) that the burst produced by SRBNs is the signal that triggers the initiation of a saccade. The high frequency burst produces a momentary release from the tonic inhibition of pontine OPNs and allows EBNs to generate a pulse of activity that is transmitted to the motoneurons. Exactly which intervening neurons are involved in converting the bursts of SRBNs into an inhibition of OPNs is unknown. Viable hypotheses exist (see e.g. Strassman et al 1986, Scudder et al 1988) but are difficult to test experimentally.

Much contemporary research is focused on the important question of how (from a vast array of potential targets) a particular target is selected as the goal for a saccadic eye movement. Current evidence suggests that information about the location of the selected target gradually develops during the reaction time interval (Sparks et al 1987) and that saccades are initiated when the time integral of cortical or collicular activity exceeds a threshold value (see Schall 2003, 2004 for reviews). The high frequency burst of collicular saccade-related burst neurons provides a precise indication of when the threshold is exceeded. But the mechanism by which activity accumulating in cortical and subcortical areas is suddenly translated into the burst of SRBNs to trigger a saccade has not been specified.

Reliability of motor signals involved in the generation of saccadic eye movements

The responses of visual neurons have been shown to be highly reproducible when the same time-varying luminance patterns are presented repeatedly (e.g. Reinagel & Reid 2000). Few comparable studies of the reliability of the motor command signals carried by individual neurons exist. In sensory studies, the physical properties of the stimulus can be held constant from trial to trial. How do we perform the motor equivalent of repeatedly presenting the same physical stimulus? If we ask a subject to make many saccades from the same initial fixation target to the same eccentric target, considerable variability in the amplitude, duration and velocity of the movement is observed (Fig. 2A). Variability in initial fixation position (centre row, left) is smaller than variability in the position reached after the primary saccade (bottom row, left). Thus, small changes in the retinal eccentricity of the target stimulus resulting from variations in initial fixation position cannot fully account for the variability in final, postsaccadic position. What are the neural sources of variability in saccade amplitude?

Methods

We examined the trial-by-trial relationship between the amplitude of saccadic eye movements and the activity of putative[1] EBNs in the paramedian pontine reticular formation (PPRF) of rhesus monkeys. These cells were chosen for study because of the strong relationships between cell activity and the parameters of the movements illustrated in Figure 1A. Action potentials were recorded from individual pontine burst neurons while monkeys made horizontal saccades to 2–5 visual targets, always starting from the same initial eye position. Initially, spikes were recorded as time stamps with 1 μs resolution using an electronic window discriminator to determine the occurrence of an action potential. The size of action potentials becomes smaller and irregular in amplitude as the burst of pontine cells progresses (Keller 1974). Thus, the window discriminator may fail to detect one or more of the smaller action potentials in the burst. Spuriously long interspike intervals would be recorded and expressed as a 'dropout' or momentary lower frequency in the instantaneous frequency records we used for the analyses presented in Figures 3–5. To avoid this potential error in measurement, we have begun recording the waveform of action potentials and checking the validity of the window discriminator output offline. These recordings have convinced us that many of the 'dropouts' are artifactual. Accordingly, the data presented in this paper are based on trials in which bursts with two or more dropouts are excluded.

Number of spikes and saccade amplitude

Figure 2B plots the number of spikes generated by a single pontine neuron during 115 saccades to one target located 20 degrees from the fixation point and 115 saccades to another target located 30 degrees from the fixation point. Because of the variation in both the amplitude and number of spikes, the data appear as two local clusters. Within each cluster, saccade amplitude varies over a 2–3 degree range and the number of spikes varies by 8–10 spikes. The correlation coefficients between number of spikes and saccade amplitude in the two clusters were 0.50 and 0.43. The distribution of correlation coefficients obtained from 36 experiments similar to the one illustrated in panel B is presented in panel C. The average correlation coefficient for these 36 data sets was 0.37.

Some of the variability in saccade amplitude observed when the animal makes repeated movements to a particular target is associated with variability in the number of spikes produced by pontine burst neurons. Under the conditions of this

---

[1] For the cells we recorded, the relationship between measures of spike activity and saccade amplitude, duration and velocity meet criteria for classifying them as EBNs. But, by definition, EBNs project monosynaptically to motoneurons and we do not know if the cells we recorded do so. For this reason, we call them putative EBNs.

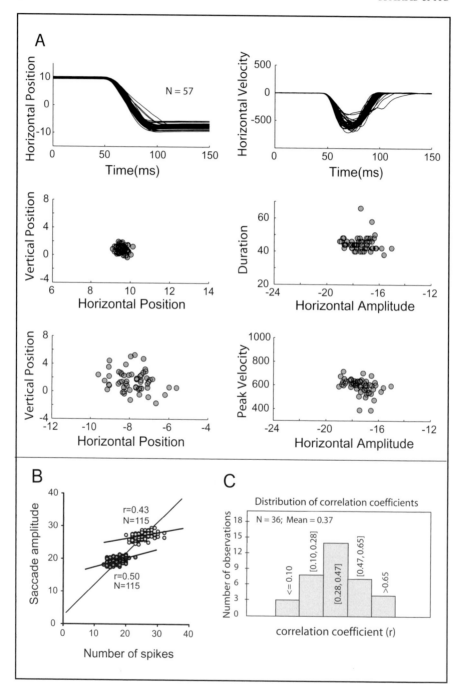

experiment, we assume that all pontine EBNs are receiving a common input specifying the same desired horizontal movement amplitude. We also assume that the actual amplitude of the executed movement is based on a summation of the output of all the cells. The correlation between an EBN's number of spikes and saccade amplitude tells us how much of the variance in saccade amplitude is explained by the activity of the neuron. The variance explained is the square of the correlation coefficient. The mean correlation coefficient of our sample is 0.37; the average proportion of variance accounted for by one neuron is about 14%. Using the rationale employed by sensory neurophysiologists when trying to estimate how many cortical cells are involved in a perceptual decision, about seven statistically independent pontine neurons like those recorded would account for the variance in saccade amplitude. However, we know that the excitability of the motoneuron pool is influenced by hundreds or thousands of pontine burst cells. We conclude, therefore, that the activity of pontine neurons is not statistically independent and that simultaneous recordings from multiple cells will reveal a high degree of correlated activity.

Variability in the profile of the burst produced by putative EBNs

Number of spikes was used as an index of trial-to-trial variability in the activity of pontine burst cells for the analysis presented in Figure 2. Number of spikes is a single global measure of a time varying process. An analysis that examines variability during the burst is presented in Figure 3. Multiple traces of eye position (P), eye velocity (V), and instantaneous spike frequency (F) are superimposed in panels A–E. The mean and standard deviation of instantaneous frequency was calculated for 2 ms bins for the bursts associated with 10, 20 and 30 degree horizontal movements. The standard deviation of instantaneous frequency is plotted in the bottom trace of panels A–E. The plots in panels A–C are aligned on movement onset and plots in panels D and E are aligned on burst onset.

FIG. 2.   Behavioural and neuronal variability during repeated saccades to the same visual target. (A) (*Left*) Plots of horizontal position (top), initial fixation position (middle) and postsaccadic position (bottom) for 57 saccades to a single target. For each, the monkey was instructed to produce a saccade 18 degrees in amplitude by looking from an initial fixation target 10 degrees to the right of the straight ahead position to an eccentric target appearing 8 degrees to the left. (*Right*) Observed variability in horizontal velocity (top), duration (middle) and peak velocity (bottom). (B) Correlation between number of spikes and saccade amplitude observed during repeated movements to two different saccade targets (see text for additional detail). The long line represents a line of best fit between number of spikes and saccade amplitude for saccades to different targets. The shorter lines represent lines of best fit obtained by a linear regression of saccade amplitude onto burst size saccade amplitude for each cluster of saccades to the same target. (C) Distribution of correlation coefficients obtained from 36 measurements similar to those illustrated in panel *B*.

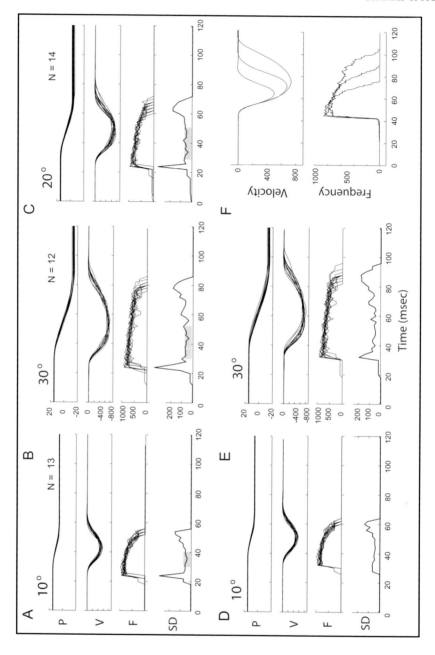

Note that the largest variability in burst activity is associated with the onset and end of the burst. During the sustained part of the burst, the standard deviation of instantaneous frequency is often less than 10% of the firing rate. The average standard deviation for the shaded regions of the plot in panel A was 48.7 spikes/s, about 6% of the average frequency (785 spikes/s) during this period (coefficient of variation = 0.06). Standard deviations and percent of average frequency for the shaded regions in panels B and C were 63.5 spikes/s (9%) and 63.2 spikes/s (9%). The variability occurring at burst onset is significantly reduced if trials are aligned on spike onset (compare A vs. D and B vs. E).

The average velocity profiles and average instantaneous frequencies for the 10, 20 and 30 degree movements shown in Panels A–C are superimposed in Panel F. The average velocity profiles follow a common trajectory for different distances, depending on saccade amplitude. For 20 and 30 degree saccades, this cell's firing rate changed almost instantaneously to the same peak frequency and then declined gradually until near the end of the burst when a sharper reduction in frequency occurs. The instantaneous frequency profiles are similar for most of the duration of the bursts associated with 20 and 30 degree movements. The variability (standard deviation) during the middle segments of the bursts associated with 20 and 30 degree movements may be comparable (panels B, C) because the cell is firing at about the same frequency during these intervals.

The data presented in Figure 4 further examine sources of variability in the burst profile of pontine burst cells. Position, velocity, instantaneous frequency and standard deviation of instantaneous frequency are plotted in panel A for 22 saccades to a target located 10 degrees from the fixation stimulus. The plots on the left were aligned on movement onset and those on the right were aligned on burst onset. Aligning the plots on burst onset significantly reduces variability of the initial segment of frequency records, but there may be a concomitant increase in variability near the end of the burst. An algorithm for sorting movements by similarity of the velocity profile was used to identify the two subsets of the movements illustrated in panels B and C. For these panels, the standard deviation plots

---

FIG. 3.   Variability in the temporal profile of EBN bursts. Plots of horizontal position (P), velocity (V), instantaneous spike frequency (F), and the standard deviation of instantaneous frequency during 13 saccades 10 degrees in amplitude (A), 1230 degree movements (B), and 1420 degree saccades (C). Instantaneous frequency is the reciprocal of interspike interval, measured by counting the number of $1\,\mu s$ clock pulses that occurred between adjacent action potentials. The plots in panels A, B and C were aligned on saccade onset. Plots aligned on the instantaneous frequency plots are shown in panels D (10 degree movements) and E (30 degree movements). Note the reduction in the initial segments of the plots of standard deviation when trials are aligned on burst onset (A vs. D; B vs. E). The average values of the standard deviations for the shaded areas in plots A, B and C are given in the text. F. Plots of the average velocity and instantaneous frequencies for the 10, 20 and 30 degree movements plotted in panels A–C.

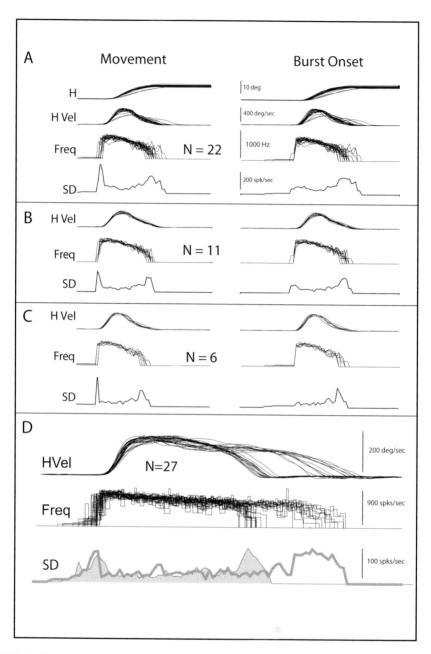

FIG. 4. Factors contributing to the variability in the burst profile. (A) Superimposed plots of horizontal position (H), velocity (HVel), instantaneous spike frequency (Freq), and the standard deviation of the instantaneous spike frequency (SD) for 22 saccades approximately 10 degrees in amplitude. Plots are aligned on saccade onset (left) or on burst onset (right). Note the reduction in the variability of instantaneous frequency during the initial segment of the burst when trials are aligned on burst onset. (B) Superimposed plots of velocity, frequency and standard deviations for a subset of 11 trials selected on the basis of the similarity of the velocity profiles. (C) Same as B but for a different subset of 6 trials. (D) Superimposed plots of horizontal velocity (first row) and instantaneous frequency (second row) for multiple saccades to visual targets located 5, 30 and 40 degrees from the fixation stimulus. Standard deviations of the instantaneous frequency of the bursts preceding the 30 degrees (shaded curve) and 40 degrees (solid line) movements. See text for more details.

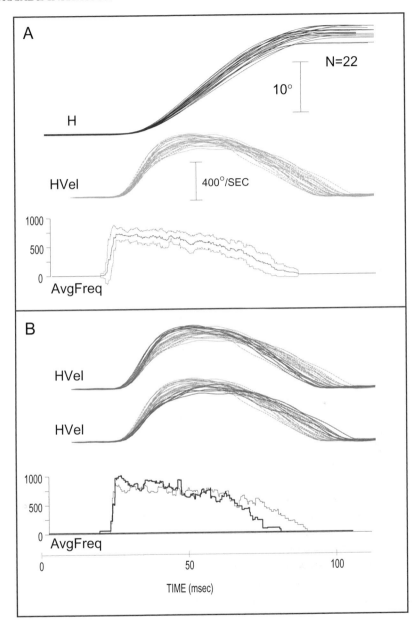

FIG. 5. Comparison of instantaneous spike frequencies for movements of the same amplitude but differing in velocity profile. (A) Superimposed plots of horizontal position (H) and velocity (HVel) for 22 saccades to the same target (20 degree eccentricity) are plotted above the average instantaneous frequency and curves showing ±1 standard deviation of 2 ms segments of the curve. (B) Two subsets of the 22 movements having different velocity profiles are highlighted in the top two plots. The bottom row shows the average instantaneous frequencies associated with each subset of movements. See text for additional details.

is a measure of variability around the mean frequency profile computed for each subset of movement, not variability around the mean for all 22 movements. One consequence of selecting a subset of movements with similar velocity profiles is a reduction in the variability of saccade duration and associated reductions in the variability of instantaneous frequency near the end of the burst. The standard deviation of instantaneous frequency during the middle segment of the burst was 8–9% of the frequency for the data presented in panel A. This was reduced to 5–6% for the subsets of movements shown in panels B and C.

The data presented in Figure 4D are consistent with the suggestion that much of the variability in the burst of pontine neurons for saccades of a given amplitude and direction occurs during the increase in firing rate at the beginning of the burst and the decrease in discharge rate near the end of the burst. Velocity and instantaneous frequency plots are superimposed for 27 movements. The area under the standard deviation plot for the 1230 degree movements is shaded to facilitate comparison with the plot of the standard deviation of the instantaneous frequency for the 1540 degree saccades. The variability in instantaneous frequency is similar during the middle, sustained portion of the bursts. Also, the increase in variability occurring near the end of the bursts is larger for the 40 degree movements. This increase in variability may correspond to the larger period over which individual burst frequencies decay to zero and this, in turn, may determine a larger range in the time when the velocity of individual movements returns to zero.

These preliminary findings lead to the following speculations. As illustrated in Figure 5A, when a subject makes repeated saccades from the same fixation target to the same eccentric target, considerable variability is observed in final position and velocity profile. The burst of putative EBNs activating the appropriate motoneuron pools also displays variability, as illustrated by the plot of average instantaneous frequency and the surrounding curves representing one standard deviation boundaries. However, if subsets of movements with distinguishably different velocity profiles are selected, the instantaneous frequency plots are also distinguishably different. The movements highlighted by darker lines in the top trace of Figure 5B reach higher peak velocities and have shorter durations than the subset of movements highlighted in the plot in the second row. The peak average instantaneous frequency is greater and burst duration is shorter for the plots associated with the subset of movements having higher peak velocity and shorter duration. Such a relationship between the activity of a single pontine burst cell and the velocity profile of a subset of movements would not be observed if other members of the large active population of EBNs had quite different temporal profiles of burst activity or if, in general, the burst profiles were heterogeneous and uncorrelated. Thus, this preliminary analysis of the variability in the burst profile of pontine neurons supports the conclusion reached based on the correlation between number of spikes and saccade amplitude—variability in the discharge of a single EBN is not independent, but strongly correlated with the activity of other active EBNs.

Further research is needed to: determine the validity of eliminating 'dropouts' from the data analysis; verify the preliminary results presented in this paper suggesting a deterministic relationship between the precise temporal patterns of firing of an EBN and the particular velocity profile of a saccade; and, ascertain to what extent the correlation between the activity of different pontine cells originates from common inputs and/or from synaptic connections between EBNs.

Acknowledgements

This research was supported by a NIH grant EY01189-32. John Maunsell made many helpful suggestions at all stages of this research. Software developed by Kathy Pearson greatly facilitated data acquisition and analysis.

References

Buttner-Ennever JA, Cohen B, Pause M, Fries W 1988 Raphe nucleus of the pons containing omnipause neurons of the oculomotor system in the monkey, and its homologue in man. J Comp Neurol 267:307–321

Curthoys IS, Markham CH, Furuya N 1984 Direct projection of pause neurons to nystagmus-related excitatory burst neurons in the cat pontine reticular formation. Exp Neurol 83:414–422

Keller EL 1974 Participation of medial pontine reticular formation in eye movement generation in monkey. J Neurophysiol 37:316–332

Keller EL 1977 Control of saccadic eye movements by midline brain stem neurons. In: Baker R, Berthoz A (eds) Control of gaze by brain stem neurons. Elsevier, Amsterdam, p 327–336

Keller EL 1979 Colliculoreticular organization in the oculomotor system. Prog Brain Res 50:725–734

King WM, Fuchs AF 1977 Neuronal activity in the mesencephalon related to vertical eye movements. In: Baker R, Berthoz A (eds) Control of gaze by brain stem neurons. Elsevier, Amsterdam, p 319–326

Moschovakis AK, Scudder CA, Highstein SM 1996 The microscopic anatomy and physiology of the mammalian saccadic system. Prog Neurobiol 50:133–254

Nakao S, Curthoys IS, Markham CH 1980 Direct inhibitory projection of pause neurons to nystagmus-related pontomedullary reticular burst neurons in the cat. Exp Brain Res 40:283–293

Reinagel P, Reid RC 2000 Precise firing events are conserved across neurons. J Neurosci 15:6837–6841

Raybourn MS, Keller EL 1977 Colliculoreticular organization in primate oculomotor system. J Neurophysiol 40:861–878

Robinson DA 1972 Eye movements evoked by collicular stimulation in the alert monkey. Vision Res 12:1795–1808

Schall JD 2003 Neural correlates of decision processes: neural and mental chronometry. Curr Opin Neurobiol 2003:182–186

Schall JD 2004 On building a bridge between brain and behavior. Annu Rev Psychol 55:23–50

Scudder CA, Fuchs AF, Langer TP 1988 Characteristics and functional identification of saccadic inhibitory burst neurons in the alert monkey. J Neurophysiol 59:1430–1454

Scudder CA, Kaneko CS, Fuchs AF 2002 The brainstem burst generator for saccadic eye movements: a modern synthesis. Exp Brain Res 142:439–462

Sparks DL 1978 Functional properties of neurons in the monkey superior colliculus: coupling of neuronal activity and saccade onset. Brain Res 156:1–16

Sparks DL 2002 The brainstem control of saccadic eye movements. Nat Rev Neurosci 3:952–964

Sparks DL, Mays LE, Porter JD 1987 Eye movements induced by pontine stimulation: Interaction with visually-triggered saccades. J Neurophysiol 58:300–318

Sparks DL, Rohrer B, Zhang Y 2000 The role of the superior colliculus in saccade initiation: a study of express saccades and the gap effect. Vision Res 40:2763–2777

Stanford TR, Freedman EG, Sparks DL 1996 The site and parameters of microstimulation determine the properties of eye movements evoked from the primate superior colliculus: evidence for independent collicular signals of saccade displacement and velocity. J Neurophysiol 76:3360–3381

Strassman A, Highstein SM, McCrea RA 1986 Anatomy and physiology of saccadic burst neurons in the alert squirrel monkey. II. Inhibitory burst neurons. J Comp Neurol 249:358–380

Strassman A, Evinger C, McCrea RA, Baker RG, Highstein SM 1987 Anatomy and physiology of intracellularly labelled omnipause neurons in the cat and squirrel monkey. Exp Brain Res 67:436–440

DISCUSSION

Schall: Have you recorded from pairs of neurons?

Sparks: No. It's on our list of things to do.

Logothetis: The EBNs you presented are the shortened ones. You would expect an input from the long-refresh neurons. What kind of variability would you expect in this previous state?

Sparks: That is an empirical question.

Logothetis: This sits between the colliculus and what you are measuring.

Sparks: From our small sample of long-lead cells, they are very much like collicular cells, except they have even larger movement fields. It is as if they are combining inputs from more than one collicular neuron. I suspect that the suggestions I made for the superior colliculus would also hold for the long-lead pontine cells. We have no data on this. I think we are going to see a lot of trial-to-trial variability, but whether or not it is going to be as highly correlated I am unsure. I don't know what to expect about correlated activity.

Derdikman: How many synapses are there from the recorded cells to the motor neuron?

Sparks: Others have intracellularly filled some of these cells and shown that they have monosynaptic connections with motor neurons. All the pontine burst cells that have similar functional properties in chronic recording experiments may not have monosynaptic connections. There can be only one synapse between these cells and the motor neuron, but there may also be multisynaptic connections.

Derdikman: What happens to the variability when you look at the motor neurons?

Sparks: Intentionally, we didn't do that. Motor neurons give bursts and a step of activity. It is hard to define the number of spikes in the burst. We decided it was easier to count the number of spikes in the burst for the cells that don't show this tonic activity afterwards.

Logothetis: Motor neurons are the ones that show the pre-emphasis: they show very high speed in the beginning and then they plateau.

Sparks: They show a pulse and step: the pulse overcomes the viscous property, and gets the eye moving, and the step overcomes the elastic properties and holds the eye in the new position.

Gold: If you look at just the end part of the saccades and relate it to this variability, have you thought about using this as a way to gain insight into the time course and quality of the integrators, the thing that produces the step response? I noticed in your traces that the time course of the variability was different in different cases. Sometimes there was more variability at the middle, sometimes nearer the end. It might be that variability in the middle is more correlated with variability at the end point, than variability at the beginning. This might tell you something about how good the integrator is at the beginning versus the middle.

Sparks: I haven't thought much about this from the integrator point of view. I expanded the time base to stretch the curves. The integrator is acting like a cumulative ISI.

Gold: How good a cumulative interspike-interval (ISI) is it? We talk about perfect integrators all the time, but we don't know how good these integrators are.

Sparks: We have been playing these bursts through the models. For a local cluster of saccades of similar amplitude, the bursts with the largest number of spikes are the slowest movements. When the cells fire with a low frequency it takes more spikes to reach the same position than it does when the burst has a higher rate. We can take data from one neuron, tweak the parameters of the existing models, and produce the proper movement. But we haven't been able to tweak the parameters of the existing models so that they will then handle multiple data sets. This is not surprising: the models are all lumped models using a single estimate of a single burst cell. It assumes an idealized burst. There needs to be a frequency sensitive element in the motor neuron–muscle interface. The way the models fail is that when we input a burst with low frequency and more spikes, the movement is too large.

Logothetis: That is consistent with what we know, because the duration of the burst is what will code for the amplitude.

Sparks: We are just beginning to do some modelling, but we haven't progressed very far.

Brecht: Do you have a sense of how many collicular burst neurons converge on your cells. You analysed the variability of saccades in relation to colliculus discharge. Do you feel there is more variability in the colliculus than in your EBN neurons, or less?

Sparks: In our old papers we say that there is not much variability, but I don't think we looked at it that carefully. We need to go back and reanalyse this. I would be surprised now if what we said in the old papers is correct! I suspect we'll find there is significant variability in collicular bursts as a function of these small variations in amplitude.

Diamond: Do you have any intuition about where the trial-to-trial variability occurs? This system ideally would be exactly identical in every trial, but it has some noise due to correlations between neurons. Or might there be some advantage in the system for visual scanning with a little bit of variance across trials? Is there any possible advantage in the system working in that way?

Sparks: There may be correlated noise and correlated signal. I think the advantages of uncorrelated noise discussed in the cortical literature may not hold at this level. I don't know of inhibitory circuits in the circuitry that would allow this to be subtracted. Uncorrelated noise could be additive at this point in the circuit, so it might not be advantageous. The correlated signal is a different issue. It could be an advantage to have correlated signals. The many burst cells are embedded in a feedback circuit. Different cells are recruited into action for different amplitude movements and the slopes of the plots of number of spikes versus saccade amplitude differ. How is the feedback signal shaped to stop each cell's activity at exactly the right time? A correlated signal could simplify the feedback problem: a subset of the cells with correlated activity could be used to shut off all the other burst cells at the same time.

Wolpert: I have been working on models of eye and arm movement in which we assume signal-dependent noise, and that amount of noise goes up with rate, for example. I am interested in your measures of standard deviation of one over the inter-spike interval. It seems you have a big component due to rate of change in motor command, at the beginning and the end. During the plateau phase, is the mean level or rate of the plateau for different amplitudes different? Do you maintain the same percentage standard deviation over those different levels?

Sparks: There are different types of cells. The ones I showed you tend to go to the same peak frequency and maintain it for varying periods. Even for these, if we look at 2 or 3 degree saccade, they will go to a lower frequency. If we do a phase plane plot of frequency versus velocity, there is a subset of cells that go to different frequencies for a small movement. This is what the prototypical cell in the literature does. In our sample of cells, the majority reach approximately the same instantaneous frequency for saccades larger than 5–7 degrees in amplitude and maintain it for different periods, depending on saccade amplitude.

Logothetis: Seven or eight years ago you introduced another complication, by showing that position in colliculus is not necessarily uniquely determining the sacchade amplitude, but the amount of activation locally, instead of causing the staircase may also cause larger amplitude sacchades. Is this correct?

Sparks: There is a site-specific maximum amplitude. If you stimulate at one locus in the superior colliculus there is a maximum amplitude that you can produce. But if you vary the duration of the stimulation you can produce any movement smaller than the site-specific maximal amplitude. The cells have to be active until the spike-specific maximum amplitude is obtained. If you terminate or truncate the input prematurely, a smaller movement occurs. Typically, the burst has sufficiently high frequency and lasts long enough that the movement is completed normally. We believe the cells are firing in a manner that the site-specific maximum amplitude normally occurs.

Logothetis: So you don't have two ways of reaching the same point.

Sparks: I hope not.

Barash: I might have got this wrong, but I had the impression that the step means that all six extraocular muscles have a positive tension all the time, and it is an agonist–antagonist relationship between them.

Sparks: I should have said something about the source of the data in the summary slide. Most of the data presented were obtained from anaesthetized cats. The muscle tensions probably aren't the same as in the alert animal.

Logothetis: I think he is asking about other types of neurons that might be involved, such as the tonic fixation neurons.

Sparks: I agree that the position of the eye depends on the ratio of innervation of the agonist–antagonist motorneuron pool. At a given position of the eye it can be associated with different levels. One motor neuron will show different firing rates for the same orbital position depending on whether this is reached by a vergence movement or a saccadic eye movement. I imagine that there will be considerable variability in motor neuron pool activity if you go to the same position repeatedly. When you go to that same position the rate of activity of the agonist–antagonist is similar.

Schall: Can you relate this to muscle fibre types and some of the heterogeneity of the population of burst neurons?

Sparks: We know very little about how the motor neurons interface with the muscle fibre types. It is an important problem that has not received the attention it deserves. We do not have good estimates of the innervation signal reaching the eye. A good estimate would involve reconstructing the pattern of activity in the agonist and antagonist motoneuron pools and assigning weights to the activity of each cell based on the properties of the muscle fibre type the cell innervates.

Multiple roles of experience in decoding the neural representation of sensory stimuli

Joshua I. Gold

University of Pennsylvania Department of Neuroscience, Philadelphia, PA 19104, USA

Abstract. Experience and perception are deeply intertwined. Experience, particularly early in life, shapes how sensory information is represented in the brain. Experience also establishes associations and can affect how sensory information guides behaviour. Central to these kinds of perceptual abilities are neural mechanisms that interpret, or decode, the brain's sensory representation, but little is known about how these decoding mechanisms depend on experience. Here I discuss several critical roles that experience might play in shaping these mechanisms. First, experience is likely to drive changes in neural connectivity to select the spatially and temporally distributed sensory signals that provide relevant information about a stimulus. Second, even the most relevant sensory signals provide incomplete information about the presence of a stimulus; also necessary is knowledge of the *a priori* probability of the stimulus, which must be learned from experience. Third, decoding noisy information is necessarily imperfect and therefore involves trade-offs like speed versus accuracy and false alarms versus misses. Experience is likely to provide ongoing feedback about the value of these trade-offs so that they might be adjusted appropriately. Each of these mechanisms appear to be capable of causing dramatic changes in sensitivity, response bias, response times and other manifestations of perceptual ability.

2005 Percept, decision, action: bridging the gaps. Wiley, Chichester (Novartis Foundation Symposium 270) p 92–107

The responses of sensory neurons to the presentation of a stimulus can be described by a likelihood distribution. This distribution gives the conditional probabilities of the possible responses, r, to a presentation of the stimulus, s; that is, $p(r|s)$. Note that, in this general case, no other assumptions are made about the stimulus or neural response (e.g. spike rate versus spike timing, labelled line versus population code, discrete versus continuous). This distribution captures the inherent variability of neural responses: repeated presentations of the same stimulus can elicit different responses, with different probabilities. Indeed, $p(r|s)$ is a complete description of how (and how well) the responses encode the stimulus. However, even with access to this description, decoding the neural responses to detect, discriminate, or identify a stimulus is a tricky proposition, because the encoded information is typically incomplete and decoding necessarily will produce errors (Schneidman et al 2003).

Recent evidence suggests that, for two-alternative decisions about sensory stimuli, the brain uses a quantity related to a ratio of likelihoods to decode incoming sensory information. The decoding scheme mirrors a theoretical framework for making decisions about uncertain information that was developed by British codebreakers during World War II and later formalized into a statistical process called the Sequential Probability Ratio Test (Gold & Shadlen 2002, Good 1979, Wald 1947). In this process, incoming information is used to compute a 'weight of evidence' based on the logarithm of the likelihood ratio (logLR) that supports or opposes the alternative interpretations. This weight of evidence is accumulated and combined with information from other sources. A decision is made when the evidence reaches a predefined criterion value. Variants of this process have been successful at explaining both accuracy and response times in a variety of psychophysical tasks (Link 1992, Ratcliff et al 1998) and appear to have correlates in the activity of single neurons involved in forming decisions that guide behaviour (Roitman & Shadlen 2002, Ratcliff et al 2003).

In this paper, I describe three challenges that the brain faces in implementing this kind of decoding scheme, each of which appears to be overcome by experience-dependent mechanisms. The first section, below, describes how the brain might read out quantities related to likelihoods without explicitly representing their distributions. The second section describes how likelihoods can be combined with information about the prior (non-conditional) probabilities of possible stimuli to determine which stimulus was presented. The third section describes the inherent trade-off between speed and accuracy in decoding ambiguous likelihoods and a possible method of balancing this trade-off by maximizing the rate of reward. Thus, experience appears to play critical roles in shaping the mechanisms that help link sensory input to motor output.

Computing likelihoods

This section describes how, under certain conditions, a simple difference in spike rates can be proportional to the logLR (Gold & Shadlen 2001, 2002). Experience is likely to select and scale the relative contributions of particular neurons to this difference, suggesting that learning might not affect the sensory representation itself but instead how that representation is read out (Dosher & Lu 1999).

The fact that sensory responses to a given stimulus are described by a distribution of likelihoods seems to imply that decoding the responses requires an explicit representation of the distribution. However, it is unclear how the brain could represent the distributions for all possible stimuli and responses. Fortunately, the responses of sensory neurons are sufficiently lawful that computing likelihood distributions seems unnecessary. Sensory neurons tend to respond selectively to a particular stimulus or stimuli, such that the preferred stimulus (e.g. an oriented bar for neurons in primary visual cortex) tends to elicit a higher rate of discharge than other stimuli. Thus, the spike rate of such a neuron can be thought of as a random variable whose expected value is greater in the presence of the preferred stimulus than otherwise. Intuitively, this property echoes likelihood: higher spike rates are more likely to be elicited when the preferred stimulus is present.

In fact, for a broad range of conditions (e.g. if the likelihood distributions are normal with equal variances, Poisson, or exponential), the neural response is a linear function of the logarithm of the ratio of likelihoods of the preferred stimulus being present (s_p) versus absent ($\sim s_p$):

$$r = k_1 + k_2 \cdot \log \frac{P(r|s_p)}{P(r|\sim s_p)}, \tag{1}$$

where k_1 and k_2 are constants (Gold & Shadlen 2001).

Interpreting the neural response as a logLR therefore involves finding an appropriate offset (k_1 in Eq. 1) and scaling factor (k_2). The offset determines which alternative (s or $\sim s$) is favoured by a given value of the response. For the logLR itself, the answer is trivial: a value >0 favours the numerator and <0 favours the denominator. What is the equivalent offset for the neural response? One solution has been suggested for decisions with two alternatives (the basic principles can be extended to decisions with many alternatives: Laming 1968, Usher & McClelland 2002). If the two alternatives are represented by neurons or pools of neurons with opposite selectivities (Fig. 1A) then a difference in activity between the two pools corresponds to an offset to 0: a value >0 favours the first alternative and <0 favours the second (Gold & Shadlen 2001, 2002).

Learning to discriminate between two stimuli therefore seems to require identifying pools of neurons selective for each of the alternatives and computing a difference in their activities. In principle, this could be achieved through a Hebbian-like mechanism that reinforces excitatory connections to a decision-maker from neurons active for one stimulus but reinforces inhibitory connections from neurons active for the other stimulus. Identifying these neural pairs is facilitated by the presence of numerous maps in cortex of parameters like location, orientation, disparity and motion direction in sensory cortex. A simple difference in activity within any of these maps can distinguish between different values of the mapped parameter. Note that either finding neurons with responses in true opposition (Fig. 1A) or scaling their responses so that they are in opposition is necessary for this scheme to generate unbiased decisions (Fig. 1B).

Interpreting a difference in activity as the logLR requires a further scaling factor (k_2 in Eq. 1). If x and y are the responses of oppositely tuned neurons and both

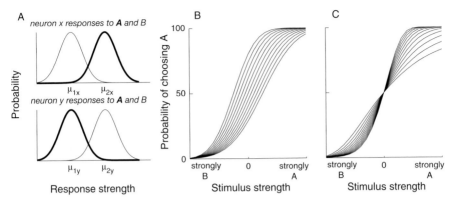

FIG. 1.   Forming decisions by approximating the logLR from a difference in sensory responses. All data are simulated. (A) Probability density functions illustrating sensory responses in opposition. The top panel shows that neuron x tends to respond more strongly to stimulus A (mean response μ_{2x}) than to B (mean μ_{1x}). The bottom panel shows that neuron y tends to respond more strongly to stimulus B (mean $\mu_{2y} = \mu_{2x}$) than to A (mean $\mu_{1y} = \mu_{1x}$). The difference $(x - y)$ approximates the logLR. (B and C) Psychometric functions describing the percentage of 'A' choices for different stimulus strengths (corresponding to different values of μ_2), calculated as the area of the distribution of $(x - y)$ that is >0. Panel B shows different curves obtained when $\mu_{2x} \neq \mu_{2y}$. Panel C shows different curves obtained by scaling the relationship between stimulus strength and μ_2.

are normally distributed with equal variances for preferred and antipreferred stimuli, then

$$\log LR_{x,y} = \frac{(\mu_1 - \mu_2)}{\sigma^2}(x - y), \tag{2}$$

where μ_1 is the expected (mean) response of each neuron to its preferred stimulus and μ_2 is the expected response of each neuron to its antipreferred stimulus (Gold & Shadlen 2001). In other words, the difference in activity $(x - y)$ is related to the logLR by a scaling factor that is proportional to the expected signal (the average difference in activity) and inversely proportional to the common noise. The bigger the signal, the more weight of evidence is provided by the difference; the bigger the noise, the less weight of evidence.

In principle, the exact value of the scaling factor is not important, because any quantity that is a monotonic function of the likelihood ratio can be used as effectively as the likelihood ratio itself in making decisions (Green & Swets 1966). In practice, however, choosing an appropriate scaling factor can be critical. For example, knowing the scaling factor makes it possible to establish a decision rule in which a fixed level of neural activity corresponds to a particular weight of evidence that, in turn, can be used to estimate expected accuracy. For such a scheme, a scaling factor that is unexpectedly large would lead to reduced accuracy (Fig. 1C) but faster response times. A scaling factor that is unexpectedly small would lead to improved accuracy (Fig. 1C) but slower response times.

Furthermore, establishing a consistent scaling between neural activity and the logLR makes it possible to combine evidence from multiple sources. For example, people are able to determine object shape using a combination of visual and haptic cues. The weight of evidence provided by each cue is scaled by its reliability, as learned from experience (Ernst & Banks 2002). In principle, such experience-dependent scaling could be used to establish a common currency to weigh evidence from different sensory stimuli along with non-sensory factors like bias, reward expectation, and utility (Glimcher 2003).

Estimating and incorporating priors

This section describes how sequential effects across trials in a perceptual task (Gilden 2001, Laming 1968) might reflect an estimate of prior probabilities. These probabilities can be combined with likelihood information to derive an accurate, probabilistic description of the sensory input.

Bayes' law quantifies the relationship between likelihoods, $p(r|s)$, and posterior probabilities, $p(s|r)$. For decisions about two possible stimuli, s_1 and s_2,

$$\log\frac{p(s_1|r)}{p(s_2|r)} = \log\frac{p(r|s_1)}{p(r|s_2)} + \log\frac{p(s_1)}{p(s_2)}, \qquad (3)$$

where the central term is the logLR and the rightmost term is the prior odds, describing the relative probabilities that s_1 or s_2 was presented independent of the sensory evidence. Eq. 3 shows that decisions based on the probability that s_1 or s_2 was present—that is, the posterior odds—require that information about the prior probabilities is combined with (in this case, added to) the information from the sensory representation (Knill & Richards 1996, Pouget et al 2003). The effect of contrast on speed judgment is one example in which perception appears to depend on a combination of likelihoods and estimates of prior probabilities (Weiss et al 2002).

Prior probabilities are, by definition, independent of the current sensory evidence and therefore must be learned from experience. Estimating the probability that a stimulus will appear seems likely to be based on the frequency with which it has already appeared. The underlying mechanisms are therefore likely to be similar to those apparent in certain foraging tasks in which subjects choose among several alternatives based on the relative frequencies with which each has been rewarded in the past (Herrnstein 1961).

A recent study examined the dynamics of this so-called 'matching' behaviour (Sugrue et al 2004). Monkeys were trained to make an eye movement to one of two visual targets, one red and one green, with different payoff frequencies associated with each target. These payoff frequencies were subject to change unexpectedly. The monkeys adopted an adaptive strategy in which their current choice appeared to be based on an estimate of the recent history of rewards, computed over ~10 trials (it seems likely that this value is itself shaped by experience; Daw & Dayan 2004).

In principle, such a strategy could be used to maintain an ongoing estimate of prior probabilities in perceptual tasks, as well. This estimate could then be combined with sensory information to guide decisions (indeed, for certain oculomotor tasks, both quantities appear to be represented in the same region of posterior parietal cortex in monkeys: Sugrue et al 2004, Roitman & Shadlen 2002). Figure 2A shows an example of a weighting function that calculates the recent history of stimulus presentations in a two-choice task. Figure 2B shows that this function can bias the subject's decisions towards more frequently presented stimuli.

Note that if such a strategy is used, a psychometric function describing performance accuracy based on stimulus strength might underestimate the subject's sensitivity, because errors attributed to insensitivity might instead be due to a bias induced by the recent sequence of trials (Fig. 2C). It is therefore useful to measure empirically the effects of recent trials on performance. One method that has been

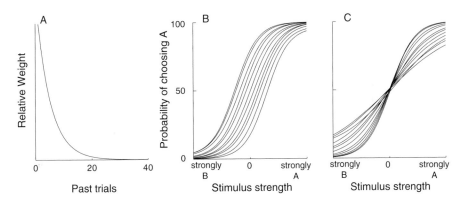

FIG. 2.   Using the recent sequence of trials to influence perceptual decisions. All data are simulated. (A) Weighting function to estimate the relative frequencies of recent stimulus presentations (the area under the curve is 1 and the two possible stimuli are coded as −1 and 1; the weighted average of a sequence of these stimuli is therefore a scalar value between −1 and 1). (B and C) Psychometric functions describing the percentage of 'A' choices for different stimulus strengths. The responses of sensory neurons were simulated as in Fig. 1. Additionally, a random sequence of stimulus presentations was generated, and each choice reflected a combination of the sensory evidence and the weighted average of recently presented stimuli. Curves are logistic fits to this simulated sequence of choices. Panel B shows biased curves that result from changing the relative frequencies with which each stimulus type was presented. Panel C shows apparent decreases in sensitivity that occur when choices become increasingly dependent on recent history.

used for response-time (RT) data involves subtracting out, for each trial, the average RT measured for all trials of that type (e.g. all trials in which a correct response to stimulus x was given), and then analysing the sequence of residuals. This method removes the main effect of stimulus on RT (e.g. stronger stimuli tend to elicit faster RTs) and then measures the extent to which variability in RTs is correlated over time (Gilden 2001). In principle, a similar method can be used to analyse sequences of choices in a two-choice task. For each stimulus condition there is a probability of making a particular choice, measured over many trials. On any given trial, the difference between this probability and unity (if the choice was made) or zero (if the alternative was chosen) is a form of residual from which the main effect of the stimulus has been removed. This sequence of residuals can then be analysed to determine the extent to which the current choice depends on the recent history of choices.

Speed–accuracy trade-off

This section describes how reward feedback can help to control the trade-off between speed and accuracy of decisions about sensory stimuli. Evidence is accumulated to a fixed threshold, the value of which can be calibrated by the recent rate of reward.

Decisions based on an accumulation over time of noisy sensory evidence face an inherent trade-off: accumulating more evidence can lead to higher accuracy but takes valuable time. One way to control this trade-off is to predefine a criterion value of evidence that determines when the accumulation should stop (Gold & Shadlen 2002, Good 1979, Wald 1947), a mechanism that the brain appears to use for at least some oculomotor tasks (Hanes & Schall 1996, Roitman & Shadlen 2002). If the accumulated evidence represents logLR and is combined appropriately with the prior odds, then the criterion value can be interpreted as the predicted odds of correctly identifying which of two alternatives is more probable, given the evidence.

Performance accuracy (Fig. 3A) and response time (Fig. 3B) are both monotonically increasing functions of the criterion value. In contrast, quantities that reflect accuracy per unit time, like the rate of reward, reach a maximum at a particular criterion value (Fig. 3C). If the value is too low, then there are too many errors and rewards are scarce. If the value is too high, then accuracy is high but time is wasted. Thus, one form of decision optimization could involve finding the criterion value that maximizes the rate of reward.

How the brain finds this optimal criterion value is unknown. In principle, this adjustment could be accomplished by a process of trial-and-error. A gradual learning process might involve raising or lowering the criterion in small increments as long as the rate of reward continues to increase. Conversely, a more rapid learning process might involve first calculating expected rates of reward from estimates of

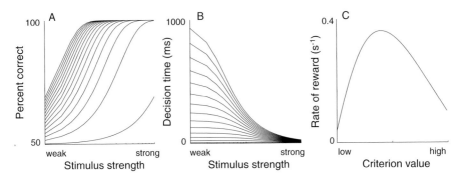

FIG. 3.   Relationship between the criterion level of evidence and reward rate. All data are simulated. Psychometric functions (A) and response times (B) for a two-choice response-time task modelled as one-dimensional Brownian motion to a pair of equidistant barriers (for details, see Link 1992 and Gold & Shadlen 2002). The different curves correspond to different barrier distances. As the barriers get closer, responses are made nearly instantaneously and performance at all stimulus strengths approaches chance. (C) The reward rate depends on both accuracy and decision time. A maximum occurs when the decision process is sufficiently accurate but does not last an excessive amount of time.

speed and accuracy for a given criterion, and then adjusting the criterion based on these estimates.

An interesting consequence of these ideas is that perceptual performance might be affected by manipulations that affect the rate of reward. For example, changing the inter-trial interval affects the number of potentially rewarded responses per unit time. Thus, subjects trying to maximize the rate of reward would have to adjust their decision criterion accordingly. In fact, preliminary studies suggest that changing the inter-trial interval in a two-choice reaction-time task causes changes in performance that are consistent with a change in the decision criterion (R. Bogacz and J. Cohen, personal communication).

Conclusion

It has been recognized for over 100 years that training can cause long-lasting improvements in the ability to detect, discriminate, or identify sensory stimuli. For example, training can improve the ability to perceive differences in visual gratings, textures, the offsets of two lines, stimulus orientation and motion direction (Gibson 1991, Goldstone 1998). The prevalence of this phenomenon implies that experience plays a central role in shaping perceptual processing in normally functioning cortex, even in adults (Karni & Bertini 1997, Buonomano & Merzenich 1998, Gilbert et al 2001). However, our understanding of the neural mechanisms that underlie perceptual learning is incomplete. Past studies have focused on plasticity in the representation of sensory information (e.g. Zohary et al 1994, Schiltz et al 1999, Crist et al 2001, Schoups et al 2001), but learning can, in principle, affect any of the series of processing stages between sensation and action that are required for performance (Graham 1989).

Mechanisms that decode the sensory representation to form decisions that guide behaviour seem particularly susceptible to the influence of experience. Selecting and scaling the appropriate sensory responses involves determining from experience their reliability in providing a weight of evidence about the sensory stimuli. Estimating prior probabilities from recent sequences of events can be combined with the weight of evidence to help make decisions about sensory stimuli. The trade-off between speed and accuracy in forming these perceptual decisions can be calibrated by an estimate of the recent rate of reward. Experience-dependent control over each of these mechanisms can, in principle, cause dramatic changes in performance accuracy, biases, and response times. Thus, these mechanisms appear to be central to the close relationship between perception and learning.

References

Buonomano DV, Merzenich MM 1998 Cortical plasticity: from synapses to maps. Annu Rev Neurosci 21:149–186

Crist RE, Li W, Gilbert CD 2001 Learning to see: experience and attention in primary visual cortex. Nat Neurosci 4:519–525

Daw ND, Dayan P 2004 Neuroscience. Matchmaking. Science 304:1753–1754

Dosher BA, Lu ZL 1999 Mechanisms of perceptual learning. Vision Res 39:3197–3221

Ernst MO, Banks MS 2002 Humans integrate visual and haptic information in a statistically optimal fashion. Nature 415:429–433

Gibson EJ 1991 An odyssey in learning and perception. In: Gibson EJ (ed) An odyssey in learning and perception. MIT Press, Cambridge, MA

Gilbert CD, Sigman M, Crist RE 2001 The neural basis of perceptual learning. Neuron 31:681–697

Gilden DL 2001 Cognitive emissions of 1/f noise. Psychol Rev 108:33–56

Glimcher PW 2003 Decisions, uncertainty, and the brain: the science of neuroeconomics. MIT Press, Cambridge, MA

Gold JI, Shadlen MN 2001 Neural computations that underlie decisions about sensory stimuli. Trends Cogn Sci 5:10–16

Gold JI, Shadlen MN 2002 Banburismus and the brain: decoding the relationship between sensory stimuli, decisions, and reward. Neuron 36:299–308

Goldstone RL 1998 Perceptual learning. Annu Rev Psychol 49:585–612

Good IJ 1979 Studies in the history of probability and statistics. XXXVI. A. M. Turing's statistical work in World War II. Biometrika 66:393–396

Graham NVS 1989 Visual pattern analyzers. Oxford University Press, Oxford

Green DM, Swets JA 1966 Signal detection theory and psychophysics. John Wiley & Sons, New York

Hanes DP, Schall JD 1996 Neural control of voluntary movement initiation. Science 274:427–430

Herrnstein RJ 1961 Relative and absolute strength of response as a function of frequency of reinforcement. J Exp Anal Behav 4:267–272

Karni A, Bertini G 1997 Learning perceptual skills: behavioral probes into adult cortical plasticity. Curr Opin Neurobiol 7:530–535

Knill DC, Richards W (eds) 1996 Perception as Bayesian inference. Cambridge: Cambridge University Press

Laming DRJ 1968 Information theory of choice reaction time. Wiley, New York

Link SW 1992 The wave theory of difference and similarity. Erlbaum, Hillsdale, NJ

Pouget A, Dayan P, Zemel RS 2003 Inference and computation with population codes. Annu Rev Neurosci 26:381–410

Ratcliff R, Rouder JN 1998 Modeling response times for two-choice decisions. Psychological Science 9:347–356

Ratcliff R, Cherian A, Segraves M 2003 A comparison of macaque behavior and superior colliculus neuronal activity to predictions from models of two-choice decisions. J Neurophysiol 90:1392–1407

Roitman JD, Shadlen MN 2002 Response of neurons in the lateral intraparietal area during a combined visual discrimination reaction time task. J Neurosci 22:9475–9489

Schiltz C, Bodart JM, Dubois S et al 1999 Neuronal mechanisms of perceptual learning: changes in human brain activity with training in orientation discrimination. Neuroimage 9:46–62

Schneidman E, Bialek W, Berry MJ, 2nd 2003 Synergy, redundancy, and independence in population codes. J Neurosci 23:11539–11553

Schoups A, Vogels R, Qian N, Orban G 2001 Practising orientation identification improves orientation coding in V1 neurons. Nature 412:549–553

Sugrue LP, Corrado GS, Newsome WT 2004 Matching behavior and the representation of value in the parietal cortex. Science 304:1782–1787

Usher M, McClelland J 2002 Hick's law in a stochastic race model with speed-accuracy tradeoff. J Math Psychol 46:704–715

Wald A 1947 Sequential analysis. Wiley, New York

Zohary E, Celebrini S, Britten KH, Newsome WT 1994 Neuronal plasticity that underlies improvement in perceptual performance. Science 263:1289–1292

DISCUSSION

Logothetis: I was uncomfortable with this comparison of single cell sensitivity and the performance using this particular design. The reason is when we are talking about the mean rate, a different way to express this is the average of the inverse of each one of the interspike intervals. This means that reliability with which you compute increases as the number of the spike increases. This can be increasing because you choose 2 s, or 5 s, or 500 ms, or because you have a lot of neurons averaging the responses. This means that if you take this statistic and use it for signal detection, instead of having two distributions being units of standard deviation apart, they are actually units of standard error apart. What is sensitive is you, not the neuron. You are becoming more sensitive because you are more reliable at computing the means. You appear to be dealing with super neurons; in reality you have a secure way of computing the mean of these neurons. I was surprised that you talked about neuronal sensitivity again.

Gold: I agree, and I wouldn't make that claim. I tried to make it clear that with this starting point it allows us to build this framework that is independent of the relationship between the sensitivity of the neuron and the behaviour. I think the computational ideas are still valid. The other answer is that I'm looking at MT responses over variable durations in a reaction time task. The plan is to do the analyses that I think you want, which is to look at different time scales and try to get a better handle on the actual timescale that the monkey is using.

Logothetis: That is right.

Gold: These data don't exist yet, so I wasn't able to talk about them.

Logothetis: If you compute the mean for the time the monkey has available to make the decision, subtracted some of the sensory delays, you might end up with periods of 150–250 ms. Within these periods you compute the mean and you put it in your signal detection.

Gold: As far as we can tell, looking at Newsome's data, it is not going to put us in a completely qualitatively different regime of reliability and sensitivity. This is why I claim that the models I am talking about are going to be valid.

Logothetis: How can you explain the behaviour given that you have the over-sensitivity of the neurons? If you do the other type of analysis you won't need to go through all these steps.

Gold: This had nothing to do with the noise and the pooling mechanism.

Treves: You said in response to an earlier question by Nikos Logothetis that you get the same kind of variability when you repeat the exact random dot stereogram. I have a vague memory of a report by Bair & Koch (1996) which showed the opposite.

Gold: They showed that there was more stereotypicality in the timing of a lot of the responses.

Logothetis: Alessandro Treves, what you may be remembering is from the work of Shadlen and Newsome themselves. In a review they showed that if you forget to update the seed of the random number generator you get identical responses. This has to be excluded before you talk about plasticity or learning, because then the local cues can not be excluded. You are not supposed to be relying on three dots that are consistently appearing in one location. There are two implications of not randomizing the seed: the first is that you may get a misguided view of what has been learned, and the other is that the neurons are actually quite sensitive.

Gold: We can find neurons that have such sensitivity, but there are plenty of neurons that don't.

Diamond: You were talking about the factors that weigh in on the accumulator. You had two sensory signals, other things such as bias and motivation, and then you added noise. You added noise only to the sensory signals. Do you think there is noise in the other inputs, such as motivation and bias? How would you put this into the equation of the accumulator reaching some threshold?

Gold: I put signal:noise ratio as being the thing that calibrates the weighing function on the sensory inputs. I put question marks on the weighing functions to the psychological variables, which acknowledge that I have no idea of how to compute it. I couldn't begin to tell you how to calibrate their relative weights. In principle, you could think of it as a reliability measure. They could well be noisy as well as the sensory systems.

Wolpert: I can see how we would tune the parameters in your model if we are getting knowledge of results, but my understanding is that in many perceptual learning tasks you get better without being told the correct answers. Also you improve with sleep. What parameters do you think improve without knowledge of results?

Gold: This isn't to discount that there are changes in the sensory representation. If you think of a sensory representation as being less vulnerable to top–down influences and shaped by lateral connections, you can imagine that repeated presentations of particular stimuli somehow reinforce these lateral connections that change the shape of the tuning function.

Wolpert: So in your experiments are they going to get rewarded for correct answers?

Gold: Yes.

Wolpert: Do the monkeys also improve without knowing the correct answer?

Gold: Correct. What is clear is that attention is necessary. It is not clear that if you don't give them a reward they are going to be attending in the same way as they do when there is a reward. This is the variable that is easier to think about.

Dehaene: The key advantage of response decision models such as the one you describe is that they can predict not just the mean but also the variability of response times. My question is, does this variability arise from the perceptual process itself, as you suggest, or from the accumulator process? In some cases, the perceptual neurons can be highly reliable, and yet there is still response variability. In my lab, Philippe Pinel, Mariano Sigman and I have studied a task where there is no perceptual variability at all, because the input is a symbol for an Arabic numeral (see in particular Sigman & Dehaene 2005). This is a number comparison task, where you have to decide whether that number is smaller or larger than a reference value. Although the stimulus is digital, there is nevertheless an effect of the numerical distance between the number and the reference, which is exactly identical to that found in more psychophysical tasks. Furthermore, variability in reaction times is still present. Thus, our current thought is that there is little variability in all of the perceptual stages before the accumulation stage and there is essentially no variability on the response side, but all of the variability comes from the process of accumulation itself.

Gold: All of these computations are carried out by neurons, so I am sure there is the potential for variability at any different stage. Part of the way of thinking I presented is to show that at least for the decision stage, every critical component depends on experience. I am sure this is true for different kinds of tasks. It is hard for me to think about generalizations across tasks if you are talking about specific details of exactly how much noise contributes to one process versus another. My guess is that it is all going to be sculpted by experience.

Dehaene: Do you have evidence that it is indeed the noise in the firing rate of overlapping perceptual representations that creates the variability in reaction time?

Gold: The analyses that exist are suggestive but flawed. There needs to be more work on this.

Haggard: I am interested in the idea that the noise may be in the accumulator. At least in the model we have seen here, the inputs to the accumulator are perceptual inputs which may have quite low noise, and a couple of other things that are known in advance, such as how biased and motivated you are. Is your observation that variation in reaction time can come from the accumulation process incompatible with the accumulator model presented here. What kind of accumulator would you need to generate high variability of non-perceptual origin?

Dehaene: Mathematically speaking I think that the models cannot be distinguished. The only mathematical assumption that is made is that the quantity that is being accumulated has a fixed value plus noise. You cannot know whether this noise comes from the percept itself, or from the process that is doing the accumulation. However, at the neuronal level, these models might perhaps be separated.

Gold: One other point is that the driving force behind the experience-dependent mechanism you are talking about was all conceived in terms of optimization processes. You want to optimise something, such as rate of reward. It is not clear to me that across all tasks you can think of the same kind of optimization as being the driving force. There are classic results in human psychophysics looking at the influence of factors such as reward magnitude on decisions in ways that relate very closely to providing different inputs. People behave very differently if you give them meaningless rewards as opposed to a meaningful award. The degree to which they optimize their computations as described by this model can vary immensely across that space. It is difficult to compare across tasks whether the degree and type of optimisation is comparable.

Schall: There are alternative models out there that we ought to keep in mind. This kind of model is based on the view that the noise is in the sensory signal that is the basis of signal detection theory, and then a single accumulation process distinguishes between the alternatives through a diffusion between the two barriers representing the two alternatives. Another class of models starts with a different assumption; biased choice theory supposes that the noise is in the decision process which leads to a race among an accumulator representing each alternative. One of the benefits of having distinct representations for each alternative was utilized in a model formulated by Vickers in which the level of activation of the non-selected item was used as a measure of the degree of ambiguity among the alternatives to improve performance in subsequent trials. The diffusion of a single accumulator has no clear representation of how much evidence favoured the non-selected alternative.

Wolpert: In the accumulator, do you have the estimate based on how long it took you to make the decision?

Schall: There are more explicit ways to get it.

Treves: You can think of fixed accumulators, or you can think of variability in the accumulator process. This is because you can change the way you accumulate evidence. For example, if there are sensory neurons that respond with limited variability to small portions of these motion patterns, but because of attention and other higher cognitive processes you amplify the outputs of some of these units during a certain trial, and of other units at another trial. This generates variability in the accumulation process which is not in the signals produced by the sensory inputs. You are just changing which units you are considering, even for the same perceptual task.

Scott: On these alternative models, couldn't you build experiments to test the difference between a horse race and one gets there first, versus a random pattern that is going one way or the other. This is simply because you put in two types of motion patterns, and you change statistically how much of the two patterns there are. You could have both rise in one situation, but as it becomes almost the same statistical probability for both you will get stuck near zero in one model and in the other you will finally have a solution.

Schall: Cognitive psychologists have got tenure in their careers by trying to distinguish diffusion and race models. Under reasonable mathematical assumptions, the two alternative formulations are indistinguishable. Behavioural testing doesn't tell them apart. Our view is that perhaps the neurons will tell them apart. There are assumptions there about which neuron maps onto which process!

Gold: Jochen has a result that begins to address the principle of this issue. Does the alternative evidence even contribute to the decision that happens? This model suggests that the underlying variable really does take into account pro-evidence and anti-evidence.

Haggard: There may be a connection here with some of the data that Shabtai Barash presented. In the free condition he had neurons which didn't win but they weren't completely losing either. They were maintaining a representation of a possible alternative, at least in that condition. This is a kind of neural evidence that would never produce a behavioural response but which shows that the second alternative is being computed.

Romo: Joshua, just for my clarification, at what moment in the learning period are you recording these cells?

Gold: We plan to record before, during and after training. The responses I showed were from before training, when the monkeys were just fixating.

Romo: So the monkeys already knew how to fixate. The monkeys are not clearly responding to the stimuli.

Gold: No. One variable that is uncontrolled now is the extent to which they are attending to the stimulus. This is the problem with the passive viewing condition. When they are performing the task we have a measure of the degree to which they are using the stimulus information to guide their choice; now they are not making any choice. Operationally, they are just fixating. But I don't know whether one monkey is just fixating and attending to the fixation point and another monkey is fixating and attending to the stimulus. In this sense I consider the result from Seidemann and Newsome that attentional modulation on the dots task was very small a good thing. This doesn't seem to directly modulate the responses of these neurons (Seidemann & Newsome 1999).

Schall: Let me pose a difficult question. In the learning study, you are recording from sampled neurons on many different days, and you have shown a change in the signalling by those neurons. How do you know you are not sampling different pools of neurons that are themselves not changing at all? Obviously, this challenge faces anyone who attempts experiments like this.

Gold: Obviously, I know I am getting different neurons. This is where Nikos Logothetis' point becomes important: we need to be able to clean up the analysis relating single neuron sensitivity to behavioural sensitivity. Whatever analysis we come up with there is going to be a relationship between neuronal sensitivity and behavioural sensitivity. The question is, how does this change over the course of learning? It will be a population argument.

Schall: Yes, but aside from that, how can we assure ourselves that the sampling is the same before and after learning?

Gold: Part of it is that I'm doing a lot of recordings before training. The big issue is the limited time window during training.

Logothetis: I don't think they ever claimed that they were recording from the same neuron.

Schall: Yes, but that is what you must do to draw the most secure conclusion about neural changes with learning.

Logothetis: The thing is, do they maintain the same conditions of statistical sampling, and do they get approximately the same type of neurons?

Gold: That is why we are doing lots of recordings before training. We have a big sample set so we are not limited to the time window during training.

Schall: Here is a set of neurons and their sensitivity is improving. I say you are getting different neurons, to which you can only say maybe, maybe not. It seems to me the rejoinder to this criticism of the sceptic is to identify a set of other, distinct neuronal characteristics by which sampling biases can be assessed.

Gold: Things like choice probability might give you a little bit of insight. The degree to which you can infer that a particular neuron is contributing to the choice the monkey makes is something that has been measured before using this kind of analysis and we plan to use in the future.

Logothetis: With the data you have there is something very simple that you can do. You can ignore the time of recording and see whether you can classify the neurons, take the two extremes, and see how consistently they correspond to the days after or days before.

Gold: I'd argue that all I have been talking about today is the attempt to overcome those ambiguities. If we have a framework for thinking about how these different neuronal mechanisms would contribute to the changes in behaviour, then we hope we'll have a more secure place to stand when we make claims that the changes we see are contributing to changes in performance.

References

Bair W, Koch C 1996 Temporal precision of spike trains in extrastriate cortex of the behaving macaque monkey. Neural Computation 8:1185–1202

Seidemann E, Newsome WT 1999 Effect of spatial attention on the responses of area MT neurons. J Neurophysiol 81:1783–1794

Sigman M, Dehaene S 2005 Parsing a cognitive task: a characterization of the mind's bottleneck. PLoS Biol 3:e37

General discussion II

Logothetis: I'm confused about how much we are learning about decision making at this meeting. Does it even make any sense to talk about decision making? This implies some sort of concrete interface somewhere and perhaps an anatomical localization. What am I missing here?

Romo: Maybe this afternoon we will learn something! I think so far we have learned about the source of variability of the neuronal responses for decision making in a particular task, and for the motor responses. It is a long-lasting problem as to whether the responses beyond the sensory areas are sensory or whether they are motor.

Derdikman: I think from Shabtai Barash's paper we learned something about decision making. The reactions of the neurons were totally different when the monkey had to make a decision versus when he was forced to perform the action.

Logothetis: My question has been misunderstood. In neuroscience there is a confusion between information and knowledge. We are in the era of information acquisition. We hear a lot of correlations and descriptions. The word 'decision making' is a nice buzz word but it has a catch: it implies you are going to say something about rules, not about correlations. The information we have is interesting and we have to accumulate it because it will mean something eventually, but it doesn't really answer what the decision making process is. This implies knowledge, which currently doesn't exist.

Barash: By knowledge do you mean of mechanisms?

Logothetis: Yes. Processes, algorithms.

Barash: I think science advances the way we are doing it: we never start from the bottom or the top, we are always in the middle. The Newsome approach has produced a lot of data which are relevant to the situation of the perceptual decisions. If you change the context of the decision, so that it is not perceptual (in our case we were careful to design it so that the perceptual load is minimal), then the temporal character of the activity changes dramatically. From the situation where the activity begins in the middle and then slowly accumulates in one direction or another, instead of this we get an asymmetrical pattern. The asymmetry is a result of the contextual situation of the decision. As far as I see it, it is a discrete situation.

Logothetis: I am talking about whether or not this is telling us something about algorithms.

Barash: In our case it is correlated to things that happened very early in the trial. The mechanisms that determine which direction is chosen are still waiting to be found.

Rizzolatti: I agree with Nikos Logothetis. The last talk concerned decision making only very marginally: it was about stimulus discrimination. The decision was already made: When you recognize stimulus A do X, when you recognize stimulus B do Y. Where is the decision mechanism?

Wolpert: Nikos Logothetis, I am not sure I understand your point. Joshua Gold has described a Bayesian mechanism that you could implement that would make decisions on the basis of things in the outside world. He has described a mechanism which could be implemented in the brain to make these decisions.

Logothetis: Let me give an example that clarifies what I said. What Bill Newsome and others are presenting is immensely interesting work. However, if we go back to 1958, there is very similar work which did exactly the same analysis with retinal ganglion cells. The results were identical. No one with a sound mind would claim that any decisions are made in the retina. There is no question that we need to do what we do. If we didn't have Newsome's work we wouldn't be able to discuss this issue. There is no question that all these correlations are useful. If you want to trivialize things, you can say that everywhere we can find cells that are selective and show different patterns.

Gold: I don't disagree with the general point that we don't know much about decisions, but one of the ways I'm trying to progress this work is as follows. The kinds of analyses that we are doing are the kinds that have been done before, but we are beginning to see the hints of broadening this framework and conceptualizing it to a more general purpose one. The framework that solves this particular problem is a general-purpose Bayesian decision-making framework. This idea of transforming a sensory representation into a weight of evidence is an important point. It says now that we have a currency that we can use to deal not just with visual motion information, but to combine all sorts of information from all sorts of different sources. How can we weigh information from multiple sources that don't seem to be otherwise related? How can we even conceptualize a computational framework for doing this? This is what I think we are doing.

Logothetis: If you want to be cynical about your data you could say that the methodology you choose is going to give this nice picture. What is 'Bayesian'? It means nothing more than saying that you are calculating a distribution given that there are some things you have learned in the past. You know the priors, and on the basis of this you calculate some posterior distributions.

Wolpert: Isn't that wonderful? It's a great framework!

Gold: We are trying to answer exactly the question you are claiming we are not even addressing. Given this general framework, what are the mechanisms that might implement it in the brain? I am not saying that we have a complete picture.

But we may be looking in a direction that allows us to begin to think about those questions.

Treves: I am not sure about the generality of the approaches we have heard about. There is no discussion that the information theoretical framework is the proper way to look at these data. There is, however, more in this approach than just that: there is always a decision between two, and only two alternatives. This is the basis for taking ratios. This seems a very specific model, applicable only to binary decision processes.

Diamond: Maybe the idea that Joshua Gold presented has been a little bit downplayed. The idea that neurons are carrying out Bayesian operations is not such a trivial idea. Once we have heard it, it seems obvious to us, but there are alternatives. Since it is not the only way that decisions could have been made it is worth considering this as a valid alternative. Is there some way that you could collect data that could distinguish between the neurons carrying out Bayesian operations, weighing alternatives, as opposed to some other model? What can you see in the data that would show that neurons are doing this?

Gold: That is the ultimate question. Part of it is refining our models so we can quantify that question. The other is coming up with the right behavioural tests. The place we are trying to gain leverage on this now is this idea of the priors. We want to talk about what is Bayesian, using prior information to update evidence. I don't have grand conclusions. So far the prior information is represented there, and we think this is in a way that is consistent with a Bayesian computation.

Diamond: Can you think of some way of measuring the state of the neuron without presenting a stimulus, to see whether the state of the neuron indicates its expectation? Are the priors in some way manifested without the stimulus being presented?

Gold: That's how I think of Bill's experience value task. This is the idea of choosing between a red and green target based on the value of what was previously experienced. There is no information from the stimulus that tells the monkey where he is going to get rewarded. It is pure prior information, based entirely on previous probabilities. We have looked at the sequence of trials in the dots task and the analysis is identical to the one that Bill is doing on this experience value task. In the dots task there is some part of the decision that is based on the stimulus.

Barash: Nikos, returning to your original question, of what have we learned today about decisions? Let me remind you of the evidence I presented about segregation of roles of parietal and prefrontal cortices, reflected in their complimentary paradoxical activities. It might give us a hint on how context affects specific processing, as in sensorimotor tasks.

Logothetis: I don't disagree with your data. I'm asking: how do you decide to go in the fast mode, for example?

Barash: Both LIP and prefrontal cortex are active all the time, maintaining representations. When the context-categorization module, possible involving prefrontal cortex, decides that this is an antisaccade trial it generates a switching command.

Logothetis: Not in your data. There were two distinct response types. In one case the visual response is much faster and also decays faster; in the other case it is delayed and has a much longer sustained portion. I see two different profiles. One gives a straight path and the other gives a complicated path.

Barash: If it is a classical visual response it is a short latency and then it goes down. There is another class of responses: these have nearly visual timings with somewhat longer latencies, but they are generated without a stimulus in the response field. These are not visual in the classic sense, and are generated only when the brain has recognized there is an antisaccade. This means that a switching signal is made available to generate this response once the monkey realizes this is an antisaccade. In order to make the decision the categorizations in the cortex must be active all the time.

Romo: It is not clear to me that the neuronal responses observed in your tests are truly sensory responses. They are simply cue responses. You can't say at the moment that these responses are encoding some features of the stimuli, because you are not varying any stimulus parameter. This is the way that we define a sensory response beyond the sensory areas. This is a permanent problem in central areas to sensory cortices: responses to a cue that occurs beyond classical sensory areas, neurons start to fire just before a movement, and it is not clear whether the responses are sensory or motion-related.

Barash: Formally, we can look at two things. First, the spatial conditions in which a neuron is firing, and the other is the timing.

Romo: You need more. You need to dissociate the sensory component from the motor component.

Barash: It's an interesting thought, although I wonder if all sensory processing should really be parametric. But you can also start by asking if the firing can be consistently classified as sensory by different criteria. In our studies, most of the activity is indeed consistently classified as visual or motor with two very different criteria.

Rizzolatti: You said that the responses are colour insensitive, but the neurons you studied receive also input from the inferior temporal cortex (IT). This input conveys also colour information.

Barash: LIP gets input from IT, but, moreover, strong 'forward' input from V4. The input is there. Many of these tests are controlled: we test them for colour. Colour was insignificant and the responses weren't different.

Schall: Over the last 10 years I believe we have made progress on a challenge voiced by David Marr in his classic book (Marr 1982). We can describe the properties of the neurons but this will never show us what they *do*. Now in this research on decision making, we are fostering a marriage of neurophysiology with cognitive psychology and formal computational models that formulate explicitly what the neurons must do and the kinds of computations they perform. Such insights have been formulated from engineering principles in the oculomotor system over the last 30 years, for example, identifying the activity of certain neurons with eye velocity and the activity of other neurons with an integration of velocity. This is a step forward.

Logothetis: There is no way that I would be saying that we are not making steps forward, or that this sort of work isn't interesting. If you are asking my opinion, I would say the analysis is still very naïve: it should perhaps be made even more complicated. We are not going to be hurt by the complications. Right now we are still talking about a single neuron: if we have 100 neurons and apply the same methodology we are going to hit the wall immediately. There are huge problems in the steps we take now. We shouldn't be confusing the progress we are making with understanding mechanisms. We are still improving our approach.

Schall: Yes, and we face still another challenge in understanding mechanism. Through computational models we can map the characteristic activity patterns of neurons to particular processes. Then the question is, how does this neuron have that characteristic? This seems to be almost impenetrable.

Logothetis: That is the challenge. One of the many things we have been doing recently is that we have been trying to use multiple tetrodes. If we do this and record from wherever, even V1, we have a diversity of responses. We have different types of spikes and a diversity of neuronal types. Some of the spikes with this technology may be coming from interneurons of different types. The tetrodes have wires $13\,\mu$m apart and we are sampling from a neighbourhood of some $50\,\mu$m. If we apply any kind of statistical analysis, we see that the same tetrodes will sometimes give cells that are more independent of each other than the responses you get between tetrodes. The reason for this is trivial: we have good isolation. This means that you are recording from one very small neighbourhood and see entirely different spike rates. This tells us that they have very distinct computational roles. There is similarity, but also there is diversity that tells us something about what and how something is being computed in the area.

Sparks: I try to think about this decision from a motor perspective. The behaviour that is being used in a lot of these experiments is the saccadic eye movement. When does a saccade occur? When the motor neurons give a burst of activity. When do the motoneurons give a burst of activity? When the EBNs generate a burst of activity. When do the EBNs burst? When the omnipause neurons (OPNs) providing tonic inhibion of EBNs turn off. So, when do the OPNs turn off? The OPNs turn off when a special class of cell in the superior colliculus (saccade-related burst neurons) gives a vigorous burst of activity. This is the trigger event: a decision has been made at this point. I did an experiment many years ago in which I varied the probability of a saccade. If the cell gives a burst the saccade occurs; if it doesn't the saccade doesn't. There's a decision. It's hard to go to mechanism, but in slice in rat superior colliculus there may be an NMDA-dependent mechanism that mediates this burst mode. When the cells go into burst mode the local neighbours also burst in synchrony, with the synchrony reducing as a function of distance from the active cell. The decision is effectively telling a particular subset of cells when to go into burst node. Decision mechanisms modulate the excitability of cells in a particular part of the motor map. My hunch is that all of these things we are manipulating, for instance by varying the size of the incentive, change the excitability of the cells.

Experimental psychologists taught us how to deal with unobservable events such as decision processes: describe them in terms of antecedent events and their consequences. The classic example is hunger. If we deprive an animal of food for so many hours or inject the animal with insulin (the antecedent conditions), the consequences are that more food will be consumed or the animal will cross a barrier of stronger shock to obtain food. It is the relationship of these antecedent events and consequences that define the unobservable event. As neuroscientists we have been forgetting this. In different sets of experiments, we manipulate the same antecedent events and measure the same consequences, but attribute the outcome to different unobservable processes (attention, motor preparation, etc.).

The fascinating part of Joshua Gold's presentation is the summing junction and the assigned weights. Two steps are needed: can we find cells that are sensitive to these variables that influence the reaction time and probability of a behaviour, that are uniquely sensitive to these, by manipulating them in operationally different ways. Then there is the interface to the trigger mechanism. We need to see if we can find neurons that are uniquely related to the hypothetical constructs that we manipulate. We also then need to understand how those cells that have those properties modify the excitability of the neurons that we know are triggering this movement. This is a motor perspective of the decision process.

Reference

Marr D 1982 Vision: a computational investigation into the human representation and processing of visual information. WH Freeman, San Francisco

Computational approaches to visual decision making

Jochen Ditterich

Center for Neuroscience, University of California, 1544 Newton Ct, Davis, CA 95616, USA

Abstract. Computational models based on diffusion processes have been proposed to account for human decision making behaviour in a variety of tasks. This study explores whether such models account for the speed and accuracy of perceptual decisions in a reaction-time random dot motion direction-discrimination task and whether they explain the decision-related activity of neurons recorded from the parietal cortex (area LIP) of monkeys performing the task. While a relatively simple diffusion model can explain the psychometric function and the mean response times, it fails to account for the response time distributions. By adding an 'urgency mechanism' to the diffusion model the psychometric function, the mean response times, and the shape of the response time distributions can be explained. Such an urgency mechanism could be implemented in different ways, but the best match between the physiological data and model predictions is provided by a diffusion process with a time-variant gain of the sensory signals. It can be shown that such a time-variant decision process allows the monkey to perform optimally (in the sense of maximizing reward rate) given the risk of aborting a trial by breaking fixation before a choice can be reported.

2005 Percept, decision, action: bridging the gaps. Wiley, Chichester (Novartis Foundation Symposium 270) p 114–128

For a couple of decades psychologists have been working on computational models to explain data patterns observed in human decision making experiments (for a review see, e.g. Ratcliff & Smith 2004). A promising class of models that seems to be able to account for data from a variety of tasks is based on bounded diffusion processes. Unfortunately, pure psychophysical studies are limited to a black box view of the system; the neural implementation of the decision making mechanism had to remain a mystery.

More recently, neurophysiologists working with awake, behaving monkeys have started studying neural mechanisms of how non-human primates solve simple decision problems (Aminoff & Goodin 1997, Britten et al 1996, Glimcher 2001, Newsome et al 1989, Platt 2002, Romo & Salinas 2001, Schall & Bichot 1998). One task which proved particularly useful for exploring neural mechanisms of primate

114

decision making is the random dot motion direction discrimination task. It has the advantage that the perception part of it (representation of motion in the visual system) and the action part of it (control of saccadic eye movements) have been well studied which allows the experimenter to focus on how the brain bridges the gap between perception and action.

Roitman & Shadlen (2002) studied a reaction time version of this task (illustrated in Fig. 1). Briefly, a monkey is watching a random dot motion stimulus on a computer screen. In this stimulus a certain fraction of the dots (called the coherence of the stimulus or the motion strength) is moving coherently in one of two possible directions. The rest of the dots are plotted at random locations in every video frame. The motion strength is varied randomly from trial to trial. The monkey watches the stimulus as long as he wants to and makes a saccade to one of two choice targets whenever he is ready, indicating the perceived direction of motion. The monkey's choice and the response time are recorded and the monkey is rewarded for a correct choice. While the monkey was performing this task Roitman and Shadlen recorded from single neurons in the lateral intraparietal area (LIP). This area has been associated with visual attention (Bisley & Goldberg 2003) and eye movement planning (Mazzoni et al 1996, Snyder et al 1998) and neurons with different properties have been described. Roitman & Shadlen (2002) recorded only from neurons with sustained activity during the memory period of a remembered saccade task. Interestingly, these neurons showed decision-related activity: early on in the trial their

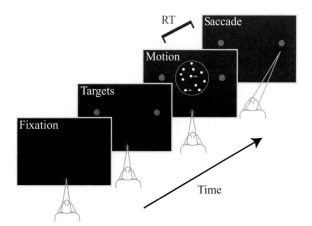

FIG. 1.   Reaction time version of the random dot motion direction discrimination task (Roitman & Shadlen 2002). A monkey watches a random dot motion pattern on a computer screen. Whenever he is ready he makes a saccade to one of two choice targets, indicating the perceived direction of motion. The monkey's choice and the response time are recorded and the monkey is rewarded for a correct choice.

response was dominated by the sensory properties of the visual stimulus, while later on in the trial their response was dominated by the monkey's choice.

This manuscript addresses the question of whether it is possible to bridge the gap between the human psychophysical studies mentioned above and the just-mentioned non-human primate study. Can a computational model based on bounded diffusion processes explain the behavioural results of the monkey study? And can it also account for the observed physiological data?

Results and discussion

The structure of the models, which will be discussed here, was motivated by several key observations made in the analysis of the LIP recordings performed by Roitman & Shadlen (2002).

First, in contrast to the response of sensory neurons in the middle temporal visual area (MT) to random dot motion patterns which is essentially flat over time (Britten et al 1993), the neurons in LIP showed a continuous and approximately linear increase or decrease of their mean firing rate as a function of time (depending on the net direction and the motion strength of the stimulus). This observation suggested that the LIP activity might reflect the temporal integral of sensory activity.

Second, the average LIP response to pure noise stimuli (no net motion) was approximately flat over time. MT neurons, however, are relatively active when a pure noise stimulus is presented (Britten et al 1993). Thus, if the LIP response reflected the temporal integral of the activity of particular sensory neurons it should have increased over time rather than being flat. The LIP activity therefore probably reflects the temporal integral over the difference in the activities of opposing sensory pools. For example, if the monkey had to discriminate between rightward and leftward motion, sensory neurons preferring leftward and rightward motion, respectively, would be approximately equally active when a pure noise stimulus is presented. The average difference between the activities of both pools would be zero and the integral therefore flat. Stimuli with net motion to the right cause 'rightward neurons' to fire stronger than 'leftward neurons'. The net evidence for rightward motion (difference between 'right' and 'left') would be positive and its integral would therefore increase over time. Similarly, stimuli with net motion to the left cause 'leftward neurons' to fire stronger than 'rightward neurons'. The net evidence for rightward motion would be negative and its integral would therefore decrease over time.

Third, the mean firing rate of LIP neurons turned out to be extremely stereotyped during the last 200 ms before the monkeys made a saccade to the target inside the response field of the recorded neuron. The motion strength of the stimulus did not have a strong influence on this activity. This observation suggests that an eye movement response might be triggered as soon as the LIP activity reaches a fixed critical level or threshold.

A time-invariant diffusion model

Putting these elements together resulted in a model with the structure shown in Fig. 2a. Two integrators collect the net evidence for a particular direction of motion (y_R and y_L, respectively) calculated as the difference between the activities of opposing sensory pools (x_R and x_L). The first integrator whose activity reaches a critical level (threshold) determines the choice and the decision time. Since y_L ($= x_L - x_R$) is just a negative copy of y_R ($= x_R - x_L$; perfect anti-correlation), the model can be mathematically formulated as a one-dimensional diffusion process with two boundaries. The fluctuating nature of the outputs of the sensory pools is captured by the assumption that y_R is described by a normal random process. Further assumptions of the model include that the mean of y_R (the drift rate of the diffusion process) is a linear function of the motion strength of the visual stimulus. The proportionality factor is the *first parameter* of the model. It is further assumed that the drift rate shows a random variation from trial to trial. As has been reported by Ratcliff & Rouder (1998), such a feature is necessary to reproduce longer response times for errors than for correct choices. The drift rate is assumed to be normally distributed and the standard deviation of this distribution is the *second parameter* of the model. The variance of the diffusion process is assumed to be constant and this is the *third parameter* of the model. The boundaries of the diffusion process are fixed and they are arbitrarily set to 1 and −1 (which can be done without loss of generality). The integration always starts at 0. The last assumption is that the response time (RT) has two additive and independent components: the decision time, which is determined by the first passage time of the diffusion process, and a component which is not decision-related and reflects other processes like motor execution. This residual time is assumed to be a constant, which is the *fourth parameter* of the model.

The model parameters were estimated by minimizing the sum of squared errors in the mean RTs. The model predictions were calculated using a numerical solution for the first passage time problem (Smith 2000). The optimal model parameters are: drift rate = 0.0194·coherence [−1 . . . 1], standard deviation of drift rate: 0.0014, variance of diffusion process: 0.0017, residual time: 322 ms.

Figure 2b shows a comparison between the experimental data and the model predictions. As can be seen on the left side, there is relatively good agreement between the psychometric functions. The model can also account for the mean response times (right side). There is, however, a major discrepancy between the shapes of the RT distributions (as shown in Fig. 2c). The model predicts heavily skewed distributions with a long exponential tail whereas the distributions observed in the experiment were much more symmetrical.

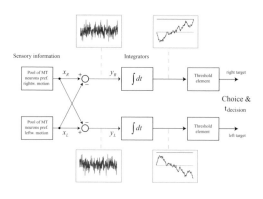

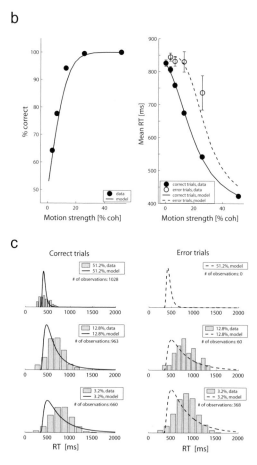

FIG. 2.   Time-invariant diffusion model. (a) Model structure. Two integrators collect the net evi-dence for a particular direction of motion (y_R and y_L, respectively) calculated as the difference between the activities of opposing sensory pools (x_R and x_L). A choice is made as soon as the activity of one of the integrators reaches a critical threshold. (b) Psychometric (left) and chrono-metric (right) functions. Comparison between data (symbols) and model predictions (lines). The error bars indicate standard errors. (c) RT distributions. Comparison between data (histograms) and model predictions (lines). The shapes do not match.

A time-variant diffusion model

The time-invariant diffusion model could account for major parts of the behavioural data. However, the RT distributions could not be explained. I will demonstrate that a time-variant diffusion model is able to account for all studied aspects
of the behavioural data (psychometric function, chronometric functions) including
the shape of the RT distributions. The basic idea is that the monkeys might want
their reward sooner rather than later, which creates some 'urgency' to finish a trial.
There are different possibilities of how such an 'urgency' mechanism could be
incorporated into the model. Here I present a model based on a time-variant gain
mechanism since it provided the best overall match with both the behavioural and
the physiological data.

Figure 3a shows the structure of the time-variant diffusion model. Like in the
time-invariant model, two integrators collect the net evidence for a particular direction of motion. The net evidence is again calculated as the difference between the
activities of opposing pools of sensory neurons, but it is now sent through a time-
variant gain stage before it enters the integrator. The gain function is plotted below
the gain elements in Figure 3a. It shows an approximately linear increase and satu-
rates after approximately 1.2 s. In contrast to the time-invariant model, the two net
evidence signals are no longer assumed to be perfectly anti-correlated. This feature
empowers the model to explain a crucial aspect of the physiological data, which
will be addressed later. The equivalent biological mechanism would be that the two
net evidence signals are derived from different subpopulations of sensory neurons.
In the model presented here, the correlation coefficient was set to −0.5. The mathematical description of this model is a two-dimensional diffusion process. The cal-
culation of the model predictions was based on a Markov chain approximation of
the bounded diffusion process. The model parameters were estimated using a
maximum likelihood procedure taking into account all aspects of the behavioural
data (distribution of choices and distribution of RTs).

As in the case of the time-invariant model, the initial drift rate was assumed to
be a linear function of the motion strength. The proportionality factor is the *first
parameter* of the model and it was determined to be 8.41×10^{-3}. In contrast to the
time-invariant model, it was not necessary to assume a trial-by-trial variability in the
drift rate. Longer error response times are a natural consequence of the time-
variance of the model. The *second parameter* is the initial variance and it was determined to be 3.07×10^{-4}. The time-variant gain function used in the model was:

$$\gamma(t) = \frac{s_y \cdot \exp(s_x \cdot (t-d))}{1 + \exp(s_x \cdot (t-d))} + \frac{1 + (1 - s_y) \cdot \exp(-s_x \cdot d)}{1 + \exp(-s_x \cdot d)}$$

The *third, fourth, and fifth parameters* are therefore d (509 ms), s_x (0.0045), and s_y (7.40;
with *t* being measured in ms). The residual time was allowed to vary randomly from

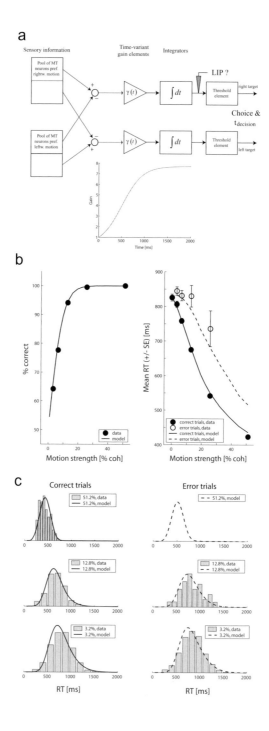

trial to trial. For simplicity, a normal distribution was assumed. The *sixth and seventh parameters* are therefore the mean residual time (251 ms) and the standard deviation of the residual time (105 ms).

Figure 3b shows that this model is able to account for the psychometric function (left side) and for the chronometric functions (right side). It is also able to account for the RT distributions (Fig. 3c).

I wondered whether in addition to be able to account for the behavioural data the model could also explain the physiological data. Since it had been suggested that the activity of the recorded neurons in area LIP represents the accumulated sensory evidence (Roitman & Shadlen 2002), I decided to simulate the model and to compare the output of the integrator (see arrow in Fig. 3a) to the recorded LIP activity. Since the model space (with an initial value of 0 and a threshold value of 1) had to be mapped to a realistic range of firing rates, I determined the best linear mapping between both spaces (minimizing the remaining sum of squared errors between data and model).

Figure 4 shows a comparison between the neural data (shown as dots) and the model predictions (shown as lines). Figure 4a shows this comparison for early on in a trial (200–350 ms after stimulus onset). The trajectories are sorted by motion strength. Dark grey indicates weak motion, light grey indicates strong motion. The actual motion strengths are shown in the figure legend. The upper part of the plot (with solid lines) represents stimuli with net motion toward the target inside the response field (RF) of the recorded neuron, the lower part (with dashed lines) represents stimuli with net motion in the opposite direction. The two dotted lines indicate the initial value (47.8 spikes/s) and the threshold value (70.2 spikes/s). Both data and model are approximately flat over time for pure noise stimuli (black), increase approximately linearly for positive motion strengths (solid lines), and decrease approximately linearly for negative motion strengths (dashed lines). The overall agreement is good, taking into account that the model has only been fitted to the behavioural data, not to the physiological data.

Figure 4b shows a comparison for later on in a trial (between 200 and 50 ms before eye movement onset). Here the trajectories are sorted by both choice (solid

FIG. 3.   Time-variant diffusion model. (a) Model structure. Two integrators collect the net evidence for a particular direction of motion calculated as the difference between the activities of opposing sensory pools. The evidence signals are sent through time-variant gain elements before integration. The gain function is plotted below the gain elements. A choice is made as soon as the activity of one of the integrators reaches a critical threshold. (b) Psychometric (left) and chronometric (right) functions. Comparison between data (symbols) and model predictions (lines). The psychometric functions match perfectly and the chronometric functions are in good agreement. (c) RT distributions. Comparison between data (histograms) and model predictions (lines). In contrast to the time-invariant model, the time-variant model can explain the RT distributions.

a

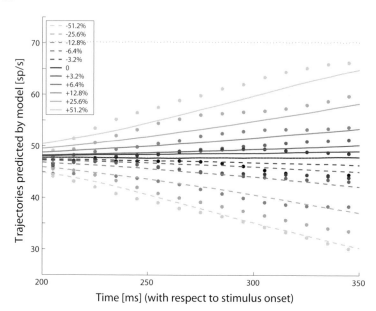

b

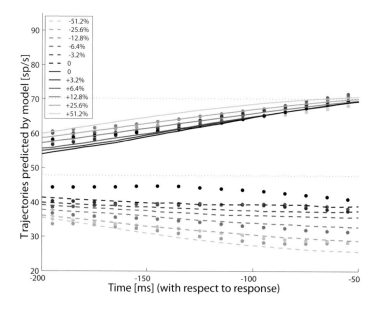

lines = target inside RF chosen, dashed lines = target outside RF chosen) and motion strength (dark grey = weak motion, light grey = strong motion), and only correct choices are shown. For both the data and the model the bundling of the trajectories is good for choices of the target inside the RF (solid lines). A critical model feature for being able to explain this observation is the assumption that an integrator stops integrating once the threshold level is reached. For choices of the target outside the RF (dashed lines) the trajectories depend much more on the motion strength. The critical model feature for being able to explain this observation is that the two evidence signals are not perfectly anti-correlated. Overall, the good agreement between data and model predictions demonstrates the power of the time-variant diffusion model in terms of being able to explain both the behavioral and the physiological data.

Reward optimization

Since a time-variant mechanism turned out to be a powerful tool for explaining the behavioural data, I wanted to explore why a time-variant decision mechanism might be desirable for the animal. One straight forward question would be whether it allowed the monkeys to increase their reward rate. I therefore calculated the reward rates, which could be obtained with a constant gain and with a time-variant gain mechanism, using relationships between response time and expected duration of a trial that were derived from the experimental data. In the experiment, the animals aborted approximately 7% of the trials by breaking fixation before they could make a valid choice. This, of course, has an influence on the reward rate, since aborting a trial means wasting time without getting rewarded. I therefore created a model with three possible outcomes: choosing either target or aborting the trial. I made sure that the model 'aborted' 7% of the trials like the monkeys did and I also matched the distributions of 'abort times'.

FIG. 4.   Time-variant diffusion model. Comparison between neural data (symbols) and model predictions (lines). (a) Early phase of the trial, aligned with stimulus onset. The trials are only sorted by motion strength (indicated in the legend). Positive values (solid lines) indicate net motion towards the target inside the RF of the recorded LIP neuron, negative values (dashed lines) indicate net motion in the opposite direction. Lighter grey means stronger motion. For both model and data the response is approximately flat for a pure noise stimulus, increases approximately linearly over time for a positive motion strength, and decreases approximately linearly for a negative motion strength. (b) Late phase of the trial, aligned with eye movement onset. The trials are sorted by choice (solid lines = target inside RF chosen, dashed lines = target outside RF chosen) and motion strength (light grey = strong motion, dark grey = weak motion). Only correct choices are shown. For both model and data the responses are nicely bundled for choices of the target inside the RF and much more dependent on motion strength for choices of the target outside the RF.

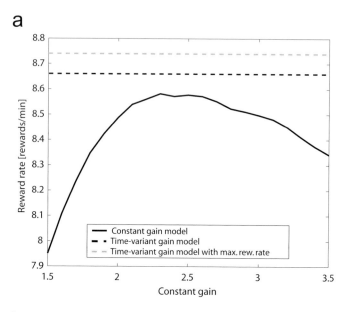

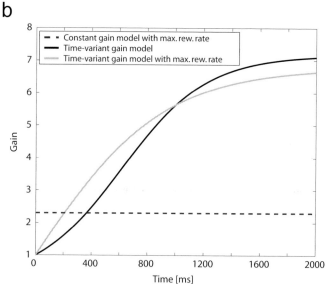

FIG. 5. Reward optimization. (a) Comparison between the reward rate associated with a constant gain model (solid line) as a function of the constant gain, the reward rate associated with a time-variant gain model yielding optimal reward (light dashed line), and the reward rate associated with the time-variant model that has been fitted to the behavioral data (dark dashed line). The comparison shows that a time-variant mechanism helps increasing the reward rate and that the time-variant model that has been fitted to the data comes close to the time-variant model yielding optimal reward. (b) Comparison between the gain functions associated with the curves in (c). The gain functions associated with the time-variant model that has been fitted to the data (dark solid line) and the time-variant model yielding optimal reward (light solid line) are very similar.

Figure 5a shows the reward rates, which could be obtained with different constant gains (solid lines). The light dashed line indicates the maximum reward rate that could be obtained with a time-variant gain function of the type used in the model discussed above. The result shows that a time-variant mechanism indeed allows for a higher reward rate compared to a time-invariant mechanism. The reward rate provided by the model that has been fitted to the behavioural data (dark dashed line) was above the maximum which could have been obtained with a time-invariant mechanism and just below the optimum. Figure 5b shows a comparison between the gain functions associated with the curves in Figure 5a. The gain function that has been determined by fitting the time-variant gain model to the data (dark solid line) is very similar to the gain function associated with the time-variant model giving maximum reward (light solid line).

Conclusions

We have seen that a diffusion model, a class of models, which has been used for explaining data from human decision making experiments, can also explain the behavioural data obtained from monkeys performing a random dot motion direction discrimination task (Roitman & Shadlen 2002). In order to be able to explain the shape of the response time distributions a time-variant mechanism was introduced into the model. This time-variant mechanism also explains why the average response time associated with wrong choices was longer than the average response time associated with correct choices. In addition to explaining the behavioural data obtained in this experiment, the model can also account for the neural activity which has been recorded from single neurons in the parietal cortex while the monkeys were performing the task.

I have further demonstrated that such a time-variant decision mechanism allows the monkeys to increase their reward rate. Furthermore, we have seen that the time-variant model resulting from fitting the behavioural data is very similar to a model resulting from optimizing the reward rate when the monkeys' risk of aborting a trial without making a valid choice is taken into account.

These observations suggest that monkeys, similar to humans, make use of decision making mechanisms that are based on the accumulation of sensory evidence up to a decision threshold. They further suggest that the nervous system is flexible enough to employ a time-variant decision mechanism if it helps the decision maker to approach optimal performance.

References

Aminoff MJ, Goodin DS 1997 The decision to make a movement: neurophysiological insights. Can J Neurol Sci 24:181–190

Bisley JW, Goldberg ME 2003 Neuronal activity in the lateral intraparietal area and spatial atten-
    tion. Science 299:81–86
Britten KH, Shadlen MN, Newsome WT, Movshon JA 1993 Responses of neurons in macaque
    MT to stochastic motion signals. Vis Neurosci 10:1157–1169
Britten KH, Newsome WT, Shadlen MN, Celebrini S, Movshon JA 1996 A relationship between
    behavioral choice and the visual responses of neurons in macaque MT. Vis Neurosci 13:87–100
Glimcher PW 2001 Making choices: the neurophysiology of visual-saccadic decision making.
    Trends Neurosci 24:654–659
Mazzoni P, Bracewell RM, Barash S, Andersen RA 1996 Motor intention activity in the macaque's
    lateral intraparietal area. I. Dissociation of motor plan from sensory memory. J Neurophysiol
    76:1439–1456
Newsome WT, Britten KH, Movshon JA 1989 Neural correlates of a perceptual decision. Nature
    341:52–54
Platt ML 2002 Neural correlates of decisions. Curr Opin Neurobiol 12:141–148
Ratcliff R, Rouder JN 1998 Modeling response times for two-choice decisions. Psychological
    Science 9:347–356
Ratcliff R, Smith PL 2004 A comparison of sequential sampling models for two-choice reaction
    time. Psychol Rev 111:333–367
Roitman JD, Shadlen MN 2002 response of neurons in the lateral intraparietal area during a com-
    bined visual discrimination reaction time task. J Neurosci 22:9475–9489
Romo R, Salinas E 2001 Touch and go: decision-making mechanisms in somatosensation. Annu
    Rev Neurosci 24:107–137
Schall JD, Bichot NP 1998 Neural correlates of visual and motor decision processes. Curr Opin
    Neurobiol 8:211–217
Smith PL 2000 Stochastic dynamic models of response time and accuracy: a foundational primer.
    J Math Psychol 44:408–463
Snyder LH, Batista AP, Andersen RA 1998 Change in motor plan, without a change in the spatial
    locus of attention, modulates activity in posterior parietal cortex. J Neurophysiol 79:2814–2819

DISCUSSION

Treves: In the model, when does the monkey select the third 'don't know'
response, breaking up the trial?

Ditterich: The way this is implemented in the model is as follows. When working
with the two-dimensional diffusion model, we have to base the calculations on an
approximation using a Markov chain process. The way the dropping out was implemented was simply by adding a third absorbing state to the Markov chain. From
every non-absorbing state you have a certain transition probability to the 'dropping
out of the trial' state. The risk of dropping out of the trial increases linearly over
time. When you do this, the distribution of 'dropping out of the trial' times matches
the one observed in the experiment.

Treves: That's another time-variant parameter.

Diamond: The time-variant gain looks really interesting. Would there be any
worthwhile test of this model by making the stimulus strength vary across the trial,
so it starts very strong and gets weaker? It would be interesting to see whether this
would affect reaction time.

Ditterich: That is a possibility. There was an experiment going on in the Shadlen Lab in which they test time-varying motion stimuli.

Schall: Do the monkeys slow down on trials after errors?

Ditterich: It is inconsistent. This is usually seen in the human data. In data from two monkeys one slowed down and the other got faster.

Gold: In one of the same monkeys, in the sequential analysis I did, they had bigger changes following correct trials than errors. Following errors the reaction time of choice tended not to be affected.

Schall: What about errors under low and high motion strength?

Gold: This didn't seem to matter.

Wolpert: I wonder whether there are different ways to think about the time-varying gain. To me it seems a bit arbitrary: you had this curve that changed over time. Could you turn it around in the sense that I have to make a decision about what is happening now based on time-distributed information and as a Bayesian estimator I know things could change. So how can I weight information I have received over the last few seconds. I should weight information I have just received more than information obtained a while ago, because things could have changed. Rather than getting into gain, effectively you are discounting old information and valuing new information more.

Ditterich: It is not identical. The discounting of old information is something you could implement by simply having a leaky integrator. It turns out that if you look at the predictions of leaky integrator models, they are not identical to what you see in the time-variant criteria or gain models.

Wolpert: If you just have a discounting factor, it seems to me that this is very similar to increasing gain over time.

Ditterich: I think it is not identical in one sense. If you think about the way the gain changes over time, it doesn't only affect the signal itself but also the noise that is on the signal. This is one of the critical aspects here that makes the difference between the time-variant gain and a leaky integrator model.

Logothetis: I was not sure I understood the gain control. Usually, pyramidal cells will go from one stage to another as a response adaptation, never a response increase over time. Do you think more things are kicking in?

Ditterich: I would say that the actual implementation of this in the neural circuit is a totally open question. It could be through the activation of more synapses, or synaptic transmission being made more efficient. The question is whether this could be done quickly enough.

Logothetis: You had two different boxes, one for left and one for right. You have the plus/minus coming to one stage, then you have the gain function and then you have the integrator. When something goes one way you say left motion, and the other way you say right motion. Is there any interaction between these two stages?

Ditterich: In the model there is no interaction later on. Of course, you could have some interaction after the gain stage or after the integrator, because all these elements are linear in the model. You can't really identify where the interaction takes place. It cannot be after the non-linear element, though, which is the threshold.

Mehler: Does your model make any prediction about what would happen if you took the tested animal and enucleated one eye? Are these areas binocularly informed?

Ditterich: I wouldn't expect any major change at all. Both eyes usually feed into these motion-sensitive areas.

Treves: The increase of gain with time could be achieved by a decrease in the standard error of the statistics. Would that be consistent with your model? That would not actually be an increasing gain in the input-output transfer function of the pyramidal cells, that itself could well adapt, i.e. lower the gain, but rather that the statistics, that is extracted from their firing, is computed more precisely as time goes on.

Ditterich: I think this wouldn't be sufficient. A better statistic would reduce noise without increasing the signal. In the end you need some kind of mechanism that drives you closer to the decision threshold, somehow. Just reducing the noise wouldn't do that.

The inferior parietal lobule: where action becomes perception

Giacomo Rizzolatti, Pier Francesco Ferrari, Stefano Rozzi and Leonardo Fogassi*

Dipartimento di Neuroscienze, Università di Parma, via Volturno 39, 43100 Parma, and
*Dipartimento di Psicologia, Università di Parma, B.go Carissimi 10, 43100 Parma, Italy

Abstract. The view defended in this article is that action and perception share the same neural substrate. To substantiate this view, the anatomical and functional organization of the inferior parietal lobule (IPL) is reviewed. In particular, it will be shown that many IPL neurons discharge selectively when the monkey executes a given motor act (e.g. grasping). Most interestingly, most of them fire only if the coded motor act is followed by a subsequent specific motor act (e.g. placing). Some of these action-constrained motor neurons have mirror properties and selectively discharge during the observation of motor acts when these are embedded in a given action (e.g. grasping for eating, but not grasping for placing). Thus, the activation of these IPL neurons allows the observer not only to recognize the observed motor act, but also to predict what will be the next motor act of the action, that is to understand the intentions of the action's agent. The finding that the same neurons that are active during the execution of specific motor acts also mediate the understanding of the 'what' and the 'why' of others' actions provides strong evidence for a common neural substrate for action and perception.

2005 Percept, decision, action: bridging the gaps. Wiley, Chichester (Novartis Foundation Symposium 270) p 129–145

It is traditionally accepted that perception derives from the association of different sensory modalities and their elaboration. Although some authors, such as Sperry (1952), found this view unsatisfactory, and stressed the importance of action in perception (see Viviani 2002), still the dominant idea is that perception and action belong to different domains (e.g. Milner & Goodale 1995).

In a recent study, Rizzolatti & Matelli (2003) re-examined the role of parietal lobe in perception and action, challenging the view that the parietal lobe is exclusively devoted to elaboration of sensory information for action. According to Rizzolatti & Matelli this notion is valid for the superior parietal lobule (SPL) and the related dorsal part the dorsal visual stream (dorso–dorsal stream). It is not valid, however, for the inferior parietal lobule (IPL) and the related ventral part of the dorsal visual stream (ventro–dorsal stream). According to these authors, IPL plays a fundamental role in perception of space and in understanding actions done by others, besides being involved in movement control. In the present essay we expand these ideas and posit that perception is not detached from action but derives from circuits originally devoted to action organization.

Anatomical organization of the inferior parietal lobule

Figure 1 shows a lateral view of the monkey brain. The posterior part of the parietal lobe consists of two lobules separated by the intraparietal sulcus: the SPL and the IPL. Several areas are buried inside the intraparietal sulcus, while others are located on the convexity. These latter areas are the main focus of the present study, and our conclusions on action–perception relationships are mostly based on their properties.

A fundamental aspect of the anatomical organization of IPL is that each of its areas receives a specific set of sensory information and is connected with a specific group of areas located in the posterior part of the frontal lobe (Rizzolatti & Luppino 2001). The most caudal IPL area, *area Opt*, receives its main visual input from area MST. Other connections of this area are with area LIP and the mesial area PGm. Frontal connections are with the dorsal part of area 46, and, to a lesser extent, with the agranular area F7. *Area PG* also receives strong visual input from area medial superior temporal (MST), but lacks the other connections of area Opt. In contrast, it is connected with somatosensory area SII and with the auditory area Tpt. Its frontal connections are with the arm-related fields of the premotor cortex, and area 46. Areas Opt and PG are often considered to be a single entity (area 7a of Brodmann). The new data reported above show that they have different efferent connections and only Opt appears to be related to the control of eye movements.

The two more rostral areas, *areas PF and PFG* are usually classified as higher-order somatosensory areas. They are indeed both targets of rich somatosensory inputs from SII, and area PF also from SI. PFG, however, receives also an important visual input from MST. Frontal connections of PFG and PF are with the ventral premotor areas F4 and F5 and the prefrontal cortex (area 46) (see also Petrides & Pandya 1984).

In addition, IPL receives an important input from the temporal lobe (Baizer et al 1991). Recent data of Luppino et al (2004) showed that all IPL areas recipient of input from MST receive input from two sectors of the temporal lobe: the rostral part of the dorsal bank of the superior temporal sulcus and, to a lesser extent, from the inferotemporal cortex.

The rostral part of the superior temporal sector is generally referred to as superior temporal polisensory area (STP). STP is the site of convergence of projections from somatosensory, auditory and visual areas. Its neuronal properties reflect this complex anatomical organization in terms of polymodal neurons and the

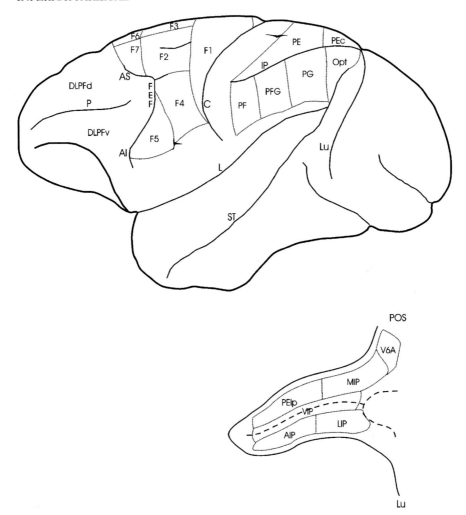

FIG. 1.   Lateral view of the monkey brain showing the parcellation of the motor and posterior parietal cortices. The areas located within the intraparietal sulcus are shown in an unfolded view of the sulcus in the lower part of the figure. For the nomenclature and definition of motor areas see Rizzolatti & Luppino (2001). Parietal areas are named according to the terminology of Pandya & Seltzer (1982), their extent is drawn, however, on the basis of recent experiments by Luppino and co-workers (Gregoriou et al 2003, Luppino 2005). AI, inferior arcuate sulcus; AS, superior arcuate sulcus; C, central sulcus; DLPFd, dorsolateral prefrontal cortex, dorsal; DLPFv, dorsolateral prefrontal cortex, ventral; L, lateral fissure; Lu, lunate sulcus; P, principal sulcus; POS, parieto-occipital sulcus; ST, superior temporal sulcus.

complexity of its visual responses (Bruce et al 1981, Perrett et al 1989, Jellema et al 2002).

The second temporal lobe input originates mostly from the fundus of the superior temporal sulcus. These areas are considered to be part of the ventral visual stream.

Functional organization of IPL

IPL is traditionally considered as an association area. Indeed, many of IPL areas receive afferents from two or more sensory channels. This multimodal association was generally considered the mechanism underlying the occurrence of cognitive activity mediated by IPL. The work of Mountcastle et al (1975) and Hyvärinen (1982) showed, however, that putting together different sensory modalities ('association function') was only one aspect of IPL organization. Another fundamental property of IPL is that of coding motor actions for the control of eye, hand and arm movement.

Hyvärinen (1981) showed that IPL, as a whole, has a somatotopic organization. This issue was recently re-investigated by Ferrari et al (2003) using simultaneous multiple electrode recording. Sensory (tactile, proprioceptive and visual) and motor maps were defined.

The motor map of IPL, as derived by this study, can be summarized as follows. The mouth motor field is located rostrally; the hand and arm motor fields are located in an intermediate position with the hand field overlapping with the mouth field rostrally and with the arm field caudally; the eye field is located caudally. At present a precise correlation between the cytoarchitectonic areas and the neuron properties is not available. Yet, preliminary reconstructions suggest that area PF coincides basically with the mouth motor field, while areas PFG and PG contain the hand and arm fields.

In agreement with previous reports (Hyvärinen 1982), IPL neurons form a functionally heterogeneous population. Most of them respond to sensory stimuli, but, most interestingly, more than 50% discharge in association with the monkey's active movements (Fogassi et al 1998).

Among the neurons responsive to sensory stimuli, about one third (33%) were activated by somatosensory stimuli, 11% by visual stimuli, and the remainder (56%) were bimodal neurons responding to somatosensory and visual stimuli (Gallese et al 2002). Note that these percentages are only indicative in the absence of a correlation with the recorded cytoarchitectonic areas.

Among neurons with visual responses ('visual neurons' and 'bimodal neurons'), a large number respond to the observations of actions done by another individual. The most represented effective observed actions are grasping, holding, manipulating and bimanual interactions. Two-thirds of these neurons become active also in association with active movements done by the monkey, typically showing congruence between their motor properties (e.g. grasping an object) and their preferred visual stimuli (observation of grasping done by the experimenter). These neurons have therefore 'mirror' properties.

What is the inferior parietal lobule for?

There are two competing views on the functional role of IPL, one stressing its role in space perception, the other in action organization. In a very influential study, Ungerleider & Mishkin (1982) proposed that the visual cortical areas in primates are organized in two information streams: one, centred on area V4, brings information on object properties to the inferotemporal lobe (ventral stream), the other centred on area MT/V5 brings information for space perception to the parietal lobe (dorsal stream). According to this view the fundamental role of the parietal lobe is perceptual and, in particular, related to space perception.

A different proposal was advanced by Milner & Goodale (Goodale & Milner 1992, Milner & Goodale 1995). On the basis of the study of a patient with visual agnosia, they re-examined the proposal of Ungerleider & Mishkin on the functional role of the two streams. They concluded that the fundamental difference between ventral and dorsal stream is not in the percept resulting from their sensory elaboration (space vs. object), but in the use that the two streams make of sensory information. The ventral stream uses visual information for perception. The dorsal stream uses visual information for the control of action (see also Jeannerod 1994).

It is clear today that both these views explain some, but not all IPL functions (see Rizzolatti & Matelli 2003). Leaving aside the motor syndromes that follow parietal damage (see De Renzi 1982), the 'space perception' hypothesis is contradicted by the motor properties of IPL neurons reviewed above. In addition, the findings of Sakata and co-workers (e.g. Sakata & Taira 1994) on area AIP strongly support the motor theory of IPL function. Neurons located in area AIP code objects in terms of their affordances and transform these affordances into the motor actions necessary to interact with objects. This sensorimotor transformation is completed by area F5 (see Jeannerod et al 1995) with which area AIP is reciprocally connected.

Conversely, there is compelling evidence also against an exclusively motor hypothesis of IPL function. First, lesions of the posterior parietal lobe in humans (the right one, in particular) produce spatial neglect (see De Renzi 1982). Second, neurons in PF, PFG and VIP code peripersonal space (Hyvärinen 1981, Colby et al 1993), and peripersonal neglect follows PF lesion in the monkey (Rizzolatti et al 1985). Third, area LIP plays a role in space coding (Andersen et al 1997, Colby & Goldberg 1999).

Considering these findings, two hypotheses can be advanced. One is that IPL is formed by several independent networks, some related to perception, others to the organization of action. The other is to consider that perception and action are functions intimately linked and that they both originate from the same networks. This last view will be defended in the next sections. For reason of space the discussion will be limited to action recognition and understanding of intentions of others. (For space perception see Rizzolatti et al 1997, Rizzolatti & Matelli 2003.)

From action control to intention understanding

Motor control in IPL

As mentioned above, many IPL neurons have motor properties. Typically, they discharge in association with specific *motor acts*. By motor acts we mean movements that have a goal, but whose goal is only partial (e.g. grasping a piece of food) and does not produce reward. We contrast the term motor act with the term *motor action*, i.e. an ensemble of motor acts that, as their final outcome, lead to a natural or conditioned reward (e.g. eating a piece of food after reaching, grasping and bringing it to the mouth).

Recently, we addressed the issue of whether IPL neurons that code a given motor act become equally active when the coded act is part of different actions. The recorded neurons were tested in two main conditions. In one the monkey reached and grasped a piece of food located in front of it and brought it to its mouth. In the second one, the monkey reached and grasped an object and placed it into a container. In the first condition the monkey ate the food, in the second the experimenter gave the monkey a reward after task accomplishment (Fogassi et al 2005).

The results showed that some neurons discharged with the same intensity regardless of the motor act that followed grasping. Most neurons, however, were influenced by the subsequent motor act (Fig. 2). The behaviour of all recorded neurons is shown in Table 1. A series of controls for grasping force, kinematics of reaching movements, and type of stimuli showed that neuron selectivity was not due to these factors.

This organization of IPL may appear to be counterintuitive. From an engineering point of view, it is probably more economical to have grasping neurons that could be used interchangeably in different actions. There is, however, a fundamental aspect of motor organization that must be taken into consideration: action fluidity. The parietal lobe organization that we just described appears to be extremely well suited for providing fluidity in action execution. The neurons coding motor acts appear to be part of pre-wired chains coding the whole action. Thus, each neuron codes a specific motor act, but simultaneously (being embedded into a specific action) is linked, and possibly facilitates, the next motor act according to the action aim.

In favour of a model assuming a facilitatory interaction between neurons forming a given chain is the organization of the receptive fields of IPL. For example, there

Unit 33

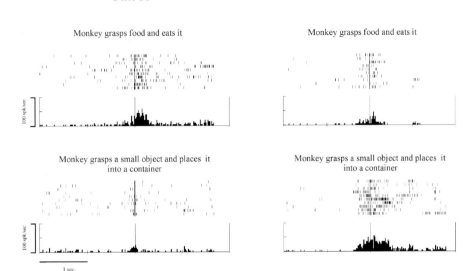

Unit 161

FIG. 2. Activity of two IPL neurons during active grasping. The neurons were tested in two conditions. In one, the monkey grasped a piece of food and ate it, in the other, the monkey grasped an object and put it into a container. Both neurons discharged during grasping, but the discharge intensity depended on the next motor act. Unit 33 discharged more strongly during grasping for eating, while unit 161 discharged more strongly during grasping for placing. Rasters and the histograms are synchronized with the moment when the monkey touched the object to be grasped. Each panel shows 10 individual trials (upper part) and the relative histogram (lower part). Abscissa: time, bin = 20 ms; Ordinate: discharge frequency.

TABLE 1   Neurons recorded during the motor task

Influenced by the final goal		*Not influenced by the final goal*	
Eating > Placing	*Placing > Eating*	*Eating = Placing*	*Total*
77 (46.7%)	29 (17.5%)	59 (35.8%)	165

are several IPL neurons that respond to passive flexion of the forearm, have tactile receptive fields located on the mouth, and discharge during mouth grasping actions. It appears therefore that tactile stimulation (as during grasping) of the hand facilitates the mouth opening through activation of a specific set of grasping neurons (L. Fogassi, P.F. Ferrari, S. Rozzi, unpublished data, Yokochi et al 2003).

Visual responses of IPL

In a series of IPL grasping mirror neurons, we tested their visual responses in the same two conditions used for studying their motor properties. The actions were performed by one of the experimenters in front of the monkey.

The results showed that the majority of IPL mirror neurons were differently activated when the observed grasping motor act was followed by bringing the food/object to the mouth or by placing it. Examples are shown in Figure 3. Table 2 summarizes the behaviour of all tested mirror neurons.

A characterizing property of all mirror neurons is the congruence of their motor and visual responses. The data just described show a further level of congruence. Mirror neurons that discharged more intensely during grasping for eating than during grasping for placing, discharged also more intensely during the observation

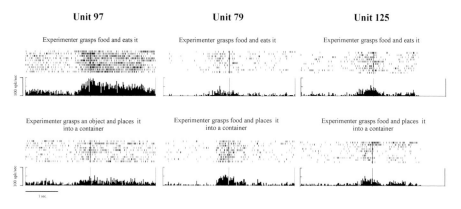

FIG. 3.   Activity of IPL mirror grasping neurons during grasping observation. The neurons were tested as explained in Fig. 2, but the motor acts were done by the experimenters. Unit 97 has a strong and prolonged discharge when the experimenter grasped a piece of food and subsequently put it into his mouth. The same grasping followed by placing was ineffective. Unit 79 has the converse behaviour. Unit 125 discharged more strongly during grasping for eating than during grasping for placing. The discharge intensity, however, in the first two trials during grasping for placing was very strong as if the presence of food, or object to be grasped, predicted eating rather than placing as the intention of the grasping individual. Conventions as in Fig. 2.

TABLE 2   Neurons recorded during the visual task

Influenced by the final goal		Not influenced by the final goal	
Eating > Placing	*Placing > Eating*	*Eating = Placing*	*Total*
23 (56.1%)	8 (19.5%)	10 (24.4%)	41

of grasping for eating. Conversely, neurons selective for grasping to place discharged strongly during the observation of this motor act.

The motor organization provides the substrate for the perception of others' intention

When an individual starts the first motor act of an action, he has clear in his mind what the final goal of the action is, to which the motor act belongs. The action *intention*, therefore, is set before the beginning of the movements and, as shown by IPL motor properties, its reflection is visible in the neuron discharge accompanying the first motor act. This intention reflection and the chained organization of IPL neurons have profound consequences on a fundamental cognitive capacity, that of understanding the intention of others.

It is generally accepted that the basic functional role of mirror neurons is that of allowing the observing individual to understand the meaning of the observed motor acts (see Rizzolatti et al 2001). The rationale of this interpretation is the following. Because the monkey knows the outcome of its motor acts, it recognizes the goal of the motor act done by another individual, when this act triggers the same set of neurons that are active during that act execution.

The surprising finding is that IPL mirror neurons, in addition to recognizing the observed motor act, are able to discriminate among identical motor acts according to the context in which they are executed. Because the discriminated motor acts are part of a chain eventually leading to the final goal of the action, this capacity allows the monkey to predict what is the goal of the observed action and, thus, to 'read' the intention of the acting individual.

The rules of this intention understanding mechanism are rather simple: if a set of grasping neurons (part of the chain 'a') fire, the observed acting individual is going to bring the food to the mouth; if, in contrast, another set of grasping neurons (part of the chain 'b') becomes active, the observed acting individual is going to put the food away.

More complex is to specify how a particular set of mirror grasping neurons is selected. After all, what the observer sees is just a hand grasping a piece of food or an object. There are various factors that might determine this selection. The first is the type of object grasped by the acting individual. Typically, monkeys grasp food in order to eat it. Thus, because of this association, the observation of an action directed towards food more likely triggers neurons coding grasping for eating than grasping for other purposes. This association is, of course, not mandatory. If it were so, the vision of a piece of food always would trigger the chain of motor acts coding eating. Thus, further factors should play a role.

One of these factors is the previous action of the observed agent. Such an interplay between the nature of the object and previous actions is illustrated in

Figure 3. Unit 125, a grasping-to-eat neuron, strongly discharged during the first trials of a sequence in which the experimenter grasped food for placing it into a container. It was as if during these initial trials, the nature of the object prevailed and made the monkey 'guess' (wrongly) that the intention of the experimenter was to eat the food. Later on, the repetition of the same action (placing) decreased the neuron's discharge and eliminated the prediction based on the nature of the stimulus.

Another fundamental factor that may influence the selection of the appropriate chain of motor acts is the context in which the first motor act is done. For example, grasping a piece of food close to a container suggests that placing would follow grasping. In contrast, food manipulation near the actor's mouth suggests that the actor intention is eating rather than placing.

Conclusions

The data presented in our study indicate a profound link between action and perception. They show that a mechanism that evolved in order to provide an efficient, fluid way to execute actions became a mechanism underlying a fundamental perceptual/cognitive faculty: understanding the intentions of others. Furthermore, the rich connections of IPL with areas coding biological action (areas of the superior temporal sulcus) and object semantics (inferotemporal lobe) render IPL particularly suitable for integrating sensory information with action representations, the mechanism that we posit as the basis for perception.

Understanding intentions of others constitutes a special domain of cognition (see Saxe et al 2004). Developmental studies show that this cognitive faculty has various components and that there are various steps through which infants acquire it. Given the complexity of the problem it would be naive to claim that the mechanism described in the present study is *the* mechanism forming the basis of intention reading. Yet, the present data indicate a neural mechanism through which an important cognitive faculty can be (in some cases) explained.

Acknowledgments

This study is dedicated to the memory of Massimo Matelli. The study was supported by EU Contract QLG3-CT-2002-00746, Mirror, EU Contract IST-2000-29689 Artesimit, by the Italian Ministero dell'Università e Ricerca, Cofin 2002, and FIRB n. RBNE01SZB4.

References

Andersen RA, Snyder LH, Bradley DC, Xing J 1997 Multimodal representation of space in the posterior parietal cortex and its use in planning movements. Annu Rev Neurosci 20:303–330
Baizer JS, Ungerleider LG, Desimone R 1991 Organization of visual inputs to the inferior temporal and posterior parietal cortex in macaques. J Neurosci 11:168–190

Bruce CJ, Desimone R, Gross CG 1981 Visual properties of neurons in a polysensory area in superior temporal sulcus of the macaque. J Neurophysiol 46:369–384

Colby CL, Goldberg ME 1999 Space and attention in parietal cortex. Annu Rev Neurosci 22:319–349

Colby CL, Duhamel J-R, Goldberg ME 1993 Ventral intraparietal area of the macaque: anatomic location and visual response properties. J Neurophysiol 69:902–914

De Renzi E 1982 Disorders of space exploration and cognition. John Wiley and Sons, Chichester, UK

Ferrari PF, Gregoriou G, Rozzi S, Pagliara S, Rizzolatti G, Fogassi L 2003 Functional organization of the inferior parietal lobule of the macaque monkey. Society for Neuroscience Abstr 919.7

Fogassi L, Gallese V, Fadiga L, Rizzolatti G 1998 Neurons responding to the sight of goal-directed hand/arm actions in the parietal area PF (7b) of the macaque monkey. Soc Neurosci Abstr 24: 257.5

Fogassi L, Ferrari PF, Gesierich B, Rozzi S, Chersi F, Rizzolatti G 2005 Parietal lobe: from action organization to intention understanding. Science 308:662–667

Gallese V, Fadiga L, Fogassi L, Rizzolatti G 2002 Action representation and the inferior parietal lobule. In: Prinz W, Hommel B (eds) Common mechanisms in perception and action: attention and performance, Vol. XIX. Oxford University Press, UK, p 334–355

Goodale MA, Milner AD 1992 Separate visual pathways for perception and action. Trends Neurosci 15:20–25

Gregoriou G, Luppino M, Matelli M 2003 The inferior parietal lobule convexity of the macaque monkey: cytoarchitectonic subdivision. Soc Neurosci Abstr 919.5

Hyvärinen J 1981 Regional distribution of functions in parietal association area 7 of the monkey. Brain Res 206:287–303

Hyvärinen J 1982 Posterior parietal lobe of the primate brain. Physiol Rev 62:1060–1129

Jeannerod M 1994 The representing brain: neural correlates of motor intention and imagery. Behav Brain Sci 17:187–245

Jeannerod M, Arbib MA, Rizzolatti G, Sakata H 1995 Grasping objects: the cortical mechanisms of visuomotor transformation. Trends Neurosci 18:314–320

Jellema T, Baker CI, Oram MW, Perrett DI 2002 Cell populations in the banks of the superior temporal sulcus of the macaque monkey and imitation. In: Melzoff AN, Prinz W (eds) The imitative mind. Development, evolution and brain bases. Cambridge University Press, Cambridge, UK, p 143–162

Luppino G 2005 Organization of posterior parietal lobe and of parieto frontal connections. In: Dehaene S, Duhamel J-R, Hauser M, Rizzolatti G (eds) From monkey brain to human brain. MIT Press, Cambridge Massachusetts p 235–252

Luppino G, Belmalih A, Calzavara R, Matelli M, Rozzi S 2004 Afferents from visual, somatosensory and auditory areas to the cortical convexity of the inferior parietal lobule of the macaque monkey. FENS Abstr. Vol 2, A025.11

Milner D, Goodale MA 1995 The visual brain in action. Oxford University Press, UK

Mountcastle VB, Lynch JC, Georgopoulos A, Sakata H, Acuna C 1975 Posterior parietal association cortex of the monkey: command functions for operations within extrapersonal space. J Neurophysiol 38:871–908

Pandya DN, Seltzer B 1982 Intrinsic connections and architectonics of posterior parietal cortex in the rhesus monkey. J Comp Neurol 204:196–210

Perrett DI, Harries MH, Bevan R et al 1989 Frameworks of analysis for the neural representation of animate objects and actions. J Exp Biol 146:87–113

Petrides M, Pandya DN 1984 Projections to the frontal cortex from the posterior parietal region in the rhesus monkey. J Comp Neurol 228:105–16

Rizzolatti G, Luppino G 2001 The cortical motor system. Neuron 31:889–901

Rizzolatti G, Matelli M 2003 Two different streams form the dorsal visual system: anatomy and functions. Exp Brain Res 153:146–157
Rizzolatti G, Gentilucci M, Matelli M 1985 Selective spatial attention: one center, one circuit or many circuits? In: Posner MI, Marin O (eds) Attention and performance. XI. Conscious and nonconscious information processing. Lawrence Erlbaum, Hillsdale NJ, p 251–265
Rizzolatti G, Fadiga L, Fogassi L, Gallese V 1997 The space around us. Science 277:190–191
Rizzolatti G, Fogassi L, Gallese V 2001 Neurophysiological mechanisms underlying the understanding and imitation of action. Nat Rev Neurosci 2:661–670
Sakata H, Taira M 1994 Parietal control of hand action. Curr Opin Neurobiol 4:847–856
Saxe R, Carey S, Kanwisher N 2004 Understanding other minds: Linking developmental psychology and functional neuroimaging. Annu Rev Psychol 55:87–124
Sperry RW 1952 Neurology and the mind-brain problem. Am Sci 40:291–312
Ungerleider LG, Mishkin, M 1982 Two visual systems. In: Ingle DJ, Goodale MA, Mansfield RJW (eds) Analysis of visual behavior. MIT Press, Cambridge, MA, p 549–586
Viviani P 2002 Motor competence in the perception of dynamic events: a tutorial. In: Prinz W, Hommel B (eds) Common mechanisms in perception and action: attention and performance, Vol. XIX. Oxford University Press, UK, p 406–442
Yokochi H, Tanaka M, Kumashiro M, Iriki A 2003 Inferior parietal somatosensory neurons coding face–hand coordination in Japanese macaques. Somatosens Motor Res 20:1–11

DISCUSSION

Porro: I would like to know more about the similarities and differences, and the reciprocal connections between the frontal and parietal mirror systems. You told us that the parietal mirror neurons code for motor acts. Do the frontal mirror neurons code for motor actions?

Rizzolatti: They also code motor acts. We believe, however, that motor act chains, similar to those we found in the inferior parietal lobule, are also present in the frontal lobe. In this sense actions are coded in both frontal lobe (area F5) and inferior parietal lobule. We need, however, to prove it.

Porro: Where does the motor action information, which shapes the response to motor acts, come from?

Rizzolatti: My guess is the prefrontal frontal lobe. In my view, the discharge coding the intention to do a certain action arrives from the prefrontal lobe and activates the neurons of a chain. It is possible that, in a chain, the first neuron only is directly activated by prefrontal input. Then, because of association between neurons forming a chain, the whole chain fires and the action is selected.

Rumiati: I have a few questions related to what we see in patients. You very cautiously said that you don't talk about imitation in relation to your monkey work. Yet, many people have been interpreting your findings as related to imitation in humans. There are some observations in patients with selective lesions who show a specific deficit in imitating actions which either have a meaning or don't have a meaning, which means actions that were already in their repertoire or were new to them. How do your findings relate to this human lesion work? My suggestion would be that it is possible that you have one type of observation if you lesion the mirror system more anteriorly and another if you lesion more posteriorly.

Rizzolatti: It is hard to explain imitation deficits in humans with monkey data, especially because monkeys do not imitate. We studied imitation in humans using fMRI (Buccino et al 2004). The task consisted in asking musically naïve participants to position the hand on a guitar neck after seeing the same chord played by a guitarist. There were three phases: (a) the participants observed the chord; (b) pause; (c) execution of the observed chord. The results showed that the circuit underlying imitation learning is formed by the inferior parietal lobule and the posterior part of the inferior frontal gyrus (the mirror neuron circuit). This circuit starts to be active during the observation phase. During the pause phase also the middle frontal gyrus (area 46) becomes active, possibly rearranging the individual motor acts coded by parietal and frontal mirror neurons. These data suggest a joint, rather than an independent activation of parietal and frontal areas in imitation.

Mehler: They were beautiful data. The problem I see is with some of the interpretations we could be tempted to make. You were talking about intentions. Somehow the intentions bring back what scientists call the frame problem. There is a nice experiment in which a researcher sees a child that imitates an experimenter to seize food on a dish with its mouth. If for some reason the experimenter has his hands occupied with something, so that now they are not available for the action, the imitation is to grasp with the hands, which is the default action. What the child is doing is actually imitating the intention.

Rizzolatti: I understand this experiment differently. In my view, this experiment shows that the child understands the goal of the experimenter ('*what* he is doing'), but not his intention ('*why* he is doing it'). Monkey mirror neurons code the goal of the action. Thus, by inference, mirror neurons are the most likely neural substrate of how the child understood the action done by the experimenter. However, the data I presented today go beyond that. They show that the mirror neuron system also codes the intention of the agent. In the experiment you mentioned the 'why' of the action is ambiguous.

Mehler: I find it very difficult to talk about a 'goal' in the absence of the mental term such as 'intention'.

Rizzolatti: It is difficult sometimes to speak with cognitive psychologists.

Tanifuji: You showed that, among various actions for the animal to observe and various actions for the animal to execute, a mirror neuron that responds to a perceived action is only activated when the animal executes that particular action. I wonder how a mirror neuron becomes sensitive to a particular set of a perceived action and action to execute. This could be explained if execution of an action and observation of the action are temporally associated, and there could be two possibilities to satisfy this condition. One possibility is that someone intentionally shows an action and at the same time one lets the animal do the same action, or vice versa. Alternatively, the actions represented by mirror neurons are limited in particular actions that naturally have temporal association. For example, eating food. In this case, when I have food, often others also have foods. In that sense, is there any specific type of actions that represented by mirror neurons?

Rizzolatti: It is possible that mirror neurons, or at least some sets of them, result from learning. A child sees an action done by himself. Since his hand is similar to that of other people's, he may generalize from actions done by his hand to actions done by the hands of others. However, this hardly explains mouth mirror neurons. They appear to be innate. This, however, should be proved experimentally.

Logothetis: I'd like to focus on some practical issues. Are there now extensive studies showing differences between these kinds of neurons and any kinds of motor neurons? You have a striking example where the monkey does something and the neuron fires, and the experimenter does the same thing and the same neuron fires in a similar way. The neuron is firing the same way whether the monkey is reaching or the experimenter is reaching. What is not firing the same way when the monkey is not actually reaching? Between this point and the real motor action there must be some interesting things happening. Either there is more activity or there is substantially less activity. Do you have these kinds of data?

Rizzolatti: Usually the response is stronger during the action done by the monkey than during the observed action.

Scott: There are some data on this. Paul Cisek and John Kalaska have done some experiments in which they have recorded in dorsal premotor during a reaching task and found that there is similar activity whether it is the experimenter reaching or the monkey (Cisek & Kalaska 2004).

Logothetis: So is the difference only in the pure motor areas? It would be interesting to have multiple electrode recordings that would tell us about neurons that may have an inhibitory role.

Rizzolatti: Now I understand better your question. The problem is why there is no action, when there is activity in premotor cortex.

Logothetis: Yes, something must be actively stopping an imitative act.

Rizzolatti: There are some data that may account for this. First, clinical data show that when there is large damage in the frontal lobe, the patients show a strong tendency to imitate others. Often this behavior is accompanied by the so-called 'utilization behaviour', a syndrome where patients compulsively grasp objects and tend to use them. These data strongly suggest that the parieto-frontal circuits that control action are, in normal individuals, tonically inhibited by frontal lobe 'centres'. Second, it has been shown that during action observation, in parallel with motor cortex excitation, there is an inhibition of motor neurons in the spinal cord (Baldissera et al 2001). Thus, simultaneously with a descending excitation (primary motor cortex is activated), there is also a descending inhibition preventing action execution.

Logothetis: Have you ever seen neurons that fired only for the observation?

Rizzolatti: Yes, in the parietal lobe. In the premotor cortex they are very rare, if any.

Krubitzer: We have been exploring that region using neuroanatomical tracing techniques and we observe that the parietal lobule has fairly dense connections with cingulate cortex. There will be some motivational component there as well. Did you do tasks in monkeys that don't involve food?

Rizzolatti: Yes, we did. Food is not crucial. Furthermore I am not sure that the connections of the cingulate cortex with the rostral part of the inferior parietal lobule are particularly rich.

Krubitzer: With area 5, which is on the rostral bank of the intraparietal sulcus, we see a fair number of connections with the cingulate cortex as well as the supplementary motor area.

Rizzolatti: Which part of the cingulate cortex?

Krubitzer: Middle and posterior. We are starting to do focal lesions in and around intraparietal sulcus, after training the monkeys to do manual tasks. We have some preliminary data with titimonkeys around the area you are referring to. Once lesioned, these monkeys can use the hand for non-intentioned movement such as locomotion. But when they try to intentionally grasp an object, they can't do it for the first two weeks. They start off by not using the hand for any behavioural task. Then they will move the hand but they won't grasp. Then they do a visual inspection of the hand and they will grasp with the non-dominant hand.

Schall: I want to explore another aspect of the word 'intention'. We can speak about intention as preparing to act, but another sense of intention surrounds whether you meant to do something you did—ownership of the action. Do mirror neurons bear any relationship to this in the sense of 'ownership'?

Haggard: There is a big confusion in the literature between 'ownership' and 'agency'. Ownership is normally used to refer to the ownership of one's own body. This might have nothing to do with action. Agency is more concerned with the question of how I recognize that I have caused a particular effect in the external world and is therefore related to action. Most of the literature has been very much concerned with the frontal planning and control of action, and has not considered the role of interpersonal mechanisms. These are two separate areas that haven't been put together except perhaps in one place: potentially they come together in a set of experiments where you perform an action and someone in the next door room is performing a similar action, and you see on screen an action being performed. Whose hand are you watching? Is it you or me? In principle you couldn't do this just with a mirror system because this would give you exactly the same information in both cases. This goes back to the question that Nikos was asking, about the nature of this magic extra area which dis-ambiguates the ambiguity inherent in the mirror system. Some authors have postulated a separate 'who' system in the brain to break this ambiguity.

Rizzolatti: I agree that the 'magic extra' is not in the mirror system, but in those areas that code the general intention to act. In normal life 'I know' when I intend to act. When I see another person acting, I have no doubt that the actions I see are not mine, but of somebody else, because of the lack of my intention to act. I would locate this 'intention to act' in the frontal lobe and in the cingulate areas. In addition, in everyday life, it is easy to disambiguate 'my actions' from those of others. Only my actions are accompanied by proprioception.

Logothetis: What you are finding implies that whatever it is that stops the action when you are observing someone else's action is as sophisticated as the neurons that carry out the action. You couldn't have anything terribly simple that doesn't understand what is going on when someone is blocking the response. To find these kinds of neurons is nice, but on the other hand it is puzzling. We would never have expected to see so much convergence at the single neuron level. It seems you have to double them in the sense that some other neuron must appreciate the observation and be aware that there is no action needed.

Diamond: Some of the neurons located in the mirror system can fire when the monkey sees something but not when it does the action. These could be inhibitory neurons or could activate inhibitory neurons, and thereby break up the linkage between seeing something and automatically doing it. But this doesn't answer the question of how they get to be so smart to know that they should inhibit the action.

Barash: There should be some switching signal, conveying whether this is an action plan to be performed, or an emulated action. What happens to this switch in the 'learn by seeing' mode? The observation and execution states have to work in synergy.

Rizzolatti: In the guitar experiment I just described, there was a delay between observation of an action and its execution, the pause phase. Of course, one can imitate also 'online'. In this case there is an immediate correspondence between the observed action and the executed action. The interesting thing about mirror neurons is that they provide neurophysiological evidence of a common substrate for perceived and performed actions.

Treves: You talked about chains of states and melodies. Have you observed any difference between the frontal and parietal mirror systems in terms of the temporal course of the responses? Is there something related to the difference between persistent activity in the temporal and frontal lobes with respect to intervening stimuli?

Rizzolatti: No. We never studied the time course of mirror neuron response as a population. Time-course in single neuron studies is a nightmare, because the variability among neurons is very high. Some discharge during the last phase of grasping and during holding, others before the hand touch the object and so on. It is very difficult to have clear results.

Hasson: In your experiment the monkey has to reach to the can in front of him, pick up an object and place it either in his mouth or in another box beside him. So the beginning of each movement, i.e. the initial reaching movement toward the can, is identical in the two conditions. Nevertheless, it seems that in your experiment the neurons in the parietal cortex are sensitive to the end goal already at the initial part. That is, if the monkey is about to place the object in his mouth only neurons that are selective for this movement will fire and vice versa. So my question is if you remove any context cue, such that the monkey cannot plan his movement in advance, what do these neurons do at the initial stage?

Rizzolatti: I suppose that only those mirror neurons that code grasping independent of the next motor act will fire.

Hasson: I think that it is a very impressive demonstration of how sensitive we are to context in real life.

Haggard: I am interested in the idea that some of these neurons are firing on the basis of the overall intention of the entire sequence. In the case of the visual response where you are observing the movements of another person, you used the word 'guess'. How do we 'guess' the intentions of other people? Presumably this is on the basis of regularities of the behaviour. Is this guess based on the regularities of our own behaviour as agents, or the regularities of the behaviour of the person we are observing? You can imagine an experiment which may be difficult in practical terms where you have the same basic component elements of the sequence, performed in one specific order by the monkey, while the same elements are put together in a different way by the experimenter that the monkey is observing. Would the monkey's mirror system generate the predictive firing on the basis of an analysis of the intentions of the observer, when the monkey is never doing the same sequence itself? It would be interesting to know whether the way in which the intention is inferred in the other is related to one's own intentions or whether one really can understand the other person's intention independently of one's own.

Rizzolatti: It is a very interesting problem, but we have no data to answer it.

Rumiati: I was thinking of the work that Dick Byrne did with apes. His claim is that they can reproduce the overall sequence, then the little things don't matter; there are individual differences. They transmit from one generation to another, from mother to child.

References

Baldissera F, Cavallari P, Craighero L, Fadiga L 2001 Modulation of spinal excitability during observation of hand actions in humans. Eur J Neurosci 13:190–194

Buccino G, Vogt S, Ritzl A et al 2004 Neural circuits underlying imitation learning of hand actions: an event-related fMRI study. Neuron 42:323–334

Cisek P, Kalaska JF 2004 Neural correlates of mental rehearsal in dorsal premotor cortex. Nature 431:993–996

The evolution of the neocortex in mammals: intrinsic and extrinsic contributions to the cortical phenotype

Sarah J. Karlen and Leah Krubitzer[1]

Department of Psychology, 174G Young Hall, University of California, One Shields Avenue, Davis, CA 95616, USA

Abstract. The neocortex is that portion of the brain that is involved in volitional motor control, perception, cognition and a number of other complex behaviours exhibited by mammals, including humans. Indeed, the increase in the size of the cortical sheet and cortical field number is one of the hallmarks of human brain evolution. Fossil records and comparative studies of the neocortex indicate that early mammalian neocortices were composed of only a few parts or cortical fields, and that in some lineages such as primates, the neocortex expanded dramatically. More significantly, the number of cortical fields increased and the connectivity between cortical fields became more complex. While we do not know the exact transformation between this type of increase in cortical field number and connectivity, and the emergence of complex behaviours like those mentioned above, we know that species that have large neocorticies with multiple parts generally have more complex behaviours, both overt and covert. Although a number of inroads have been made into understanding how neurons in the neocortex respond to a variety of stimuli, the micro and macro circuitry of particular neocortical fields, and the molecular developmental events that construct current organization, very little is known about how more cortical fields are added in evolution. In particular, we do not know the rules of change, nor the constraints imposed on evolving nervous systems that dictate the particular phenotype that will ultimately emerge. One reason why these issues are unresolved is that the brain is a compromise between existing genetic constraints and the need to adapt. Thus, the functions that the brain generates are absolutely imperfect, although functionally optimized. This makes it very difficult to determine the rules of construction, to generate viable computational models of brain evolution, and to predict the direction of changes that may occur over time. Despite these obstacles, it is still possible to study the evolution of the neocortex. One way is to study the products of the evolutionary process—extant mammal brains—and to make inferences about the process. The second way to study brain evolution is to examine the developmental mechanisms that give rise to complex brains. We have begun to test our theories regarding cortical evolution, generated from comparative studies, by 'tweaking' in a developing nervous system what we believe is naturally being modified in evolution. Our goals are to identify the constraints imposed on the evolving neocortex, to disentangle the genetic and activity dependent mechanisms that give rise to complex brains, and ultimately to produce a cortical phenotype that is consistent with what would naturally occur in evolution.

---

[1] This paper was presented at the symposium by Leah Krubitzer, to whom correspondence should be addressed.

2005 Percept, decision, action: bridging the gaps. Wiley, Chichester (Novartis Foundation Symposium 270) p 146–163

Throughout evolution, one of the most dramatic changes to the mammalian brain has been an increase in the size of the neocortex and a change in the number of cortical fields. A cortical field can be defined using a variety of criteria including architectonic appearance, neuronal response properties, and cortical and subcortical connections (Kaas 1982, 1983, Krubitzer 1995). While all mammals have cortical fields that are uniquely interconnected to form processing networks, different species have different numbers of cortical areas, and this variability is thought to generate the behavioural diversity exhibited by various mammals. In general, mammals with larger neocortices and a greater number of cortical fields appear to exhibit more complex behaviours and to possess a greater number of more flexible behavioural repertoires. Although the exact transformation between the addition of cortical fields and the observed changes in sensory, perceptual and cognitive behaviours is not known, the addition of cortical fields may act to enhance particular stimulus features, generate probabilities based on sensory experience, and construct a species-specific interpretation of the environment based on the physical parameters that a particular animal can detect (Krubitzer & Kahn 2003).

The goal of our laboratory is to understand how changes in brain size and complexity are generated in different lineages and, once generated, how these changes are translated into complex behaviours, such as perception and cognition. Specifically, we are interested in how functional areas are specified, how cortical fields are added, and how connections between fields are modified in different lineages.

Unfortunately, understanding the process of brain evolution is hindered by two major obstacles. First, changes that occur in the neocortex accumulate slowly over many generations in different lineages. As a result, cortical evolution cannot be studied directly and is not particularly amenable to laboratory experimentation. Second, unlike portions of the skeleton, soft tissue, such as the brain, is not preserved in the fossil record; therefore information regarding changes that occur in the brain is derived from endocasts of fossil skulls (Jerison 1973, see Kaas 2005 for review), which can only provide information about the size and shape of the brains of our mammalian ancestors. Because of these problems associated with studying evolution directly, one can only make inferences about the evolutionary processes. This can be done by examining the brains of living mammals and performing a comparative analysis, and by utilizing a developmental approach to examine the mechanisms that may have been altered in evolution to account for the neocortical modifications observed in different lineages. The latter approach is a viable means for understanding evolution because the evolution of the neocortex is, in essence, the evolution of developmental mechanisms that recreate brain phenotypes in successive generations and that give rise to brain changes within and across lineages over time.

What have we learned about cortical evolution using the comparative approach?

The comparative approach is a method that allows us to deduce general character-istics of the nervous system, the types of brain changes that are possible, and the constraints that direct the course of evolution (Bullock 1984, Krubitzer 2002). Using this approach, we and others (Campos & Welker 1976, Catania 2002, Johnson et al 1994, Kaas 2005, Krubitzer 1995, Krubitzer & Kahn 2003, Levitt & Eagleson 2000, Reep et al 1989) have examined a variety of species using the criteria described above and have come to some firm conclusions regarding homologous cortical areas and general features of cortical organization that all mammals share. For example, all mammals have a similar constellation of specifically interconnected cortical fields (Krubitzer 1995, Krubitzer & Huffman 2000). As shown in Figure 1, the primary auditory area, A1 (Ehret 1997), the primary somatosensory area, S1 (Johnson 1990, Kaas 1983), and the primary visual area, V1 (Rosa & Krubitzer 1999), have been identified in all, or nearly all, mammals examined (Krubitzer 1995, Krubitzer & Kahn 2003 for review; Fig. 1 phylogeny based on Murphy et al 2004). These primary areas contain a complete representation of the sensory epithelium that is coextensive with a unique architectonic appearance and pattern of connec-tivity. While the second auditory area (A2), the second somatosensory area/parietal ventral area (S2/PV), a rostral deep field (R), the second visual area (V2), and primary motor area (M1) appear to be common to all mammals as well, other cor-tical areas that have been described appear to be derived and limited to particular lineages, such as the extrastriate visual areas in primates. Regardless of the mor-phological and behavioural specializations of many mammals (e.g. Catania 2000, Henry et al 2005, Krubitzer 1995), the conserved constellation of fields shown in Figure 1 is always present, even in the absence of apparent use (Bronchti et al 2002, Cooper et al 1993, Heil et al 1991). The ubiquity of these fields, aspects of their corticocortical and thalamocortical connectivity, and their general geographic arrangement across species indicate that they were present in the common ances-tor, cannot be eliminated under most or all circumstances, and reflect the constraints imposed upon the evolving neocortex.

Despite the high degree of similarities between species, comparative studies also indicate that there are large degrees of freedom for phenotypic change. For example, the amount of cortex allotted to different sensory systems, termed sensory domains, varies across mammals and is related to the sensory receptor arrays and the senses that are most behaviourally relevant to the animal (Krubitzer & Kahn 2003). Furthermore, within a particular sensory domain, the relative size of primary

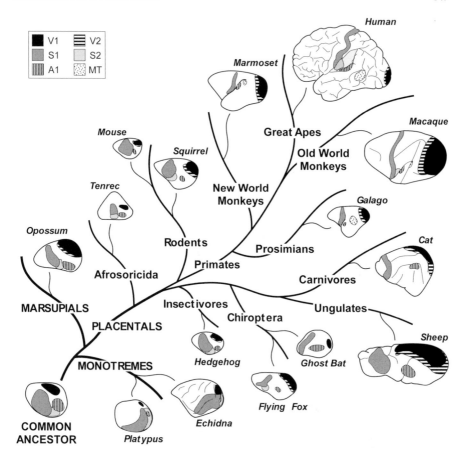

FIG. 1. An evolutionary tree depicting the phylogenetic relationship of some of the major mammalian orders and the cortical organization of some of the sensory fields that have been described in particular species. All of the mammals shown have a similar constellation of cortical fields, including A1, S1, S2, V1 and V2, as defined by architectonic appearance, neuronal response properties, and cortical and subcortical connections. The ubiquity of these fields suggests that they are most likely homologous areas that arose from a common ancestor. Other areas, such as MT, which has only been observed in the primate order, are derived and limited to particular lineages. A1, primary auditory area; S1, primary somatosensory area; S2, secondary somatosensory area; V1, primary visual area; V2, secondary visual area; MT, mediotemporal area. Phylogenetic relationships based on Murphy et al (2004). Rostral is left, medial is up.

cortical areas also varies, depending on the importance of the sensory system in question. For instance, in the arboreal squirrel, a highly visual rodent, a large proportion of the neocortex is devoted to the visual system, and the relative size of V1 is large, as compared to other primary sensory areas (Fig. 2A, based on Kaas et al 1989, Krubitzer et al 1986, Luethke et al 1988). In the mouse, which relies more

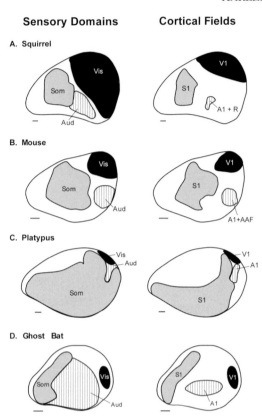

FIG. 2. The sensory domain allocation (left) and location and size of primary cortical fields (right) in the neocortex of four different mammals with different sensory specializations. The squirrel (A) is a highly visual rodent with much of its neocortex devoted to the visual system (A, left), and the relative size of V1, as compared to other primary sensory areas, is large (A, right). The mouse (B), which relies more on its somatosensory system, particularly its vibrissae, than its visual system, has a large portion of its neocortex devoted to somatosensory processing (B, left), and the relative size of S1 is larger than V1 or A1 (B, right). The duck-billed platypus (C) has an extremely well developed bill which it uses almost exclusively for feeding and navigating, and most of its neocortex is devoted to the somatosensory system (C, left). The relative size of S1, as compared to other primary fields, is quite large (C, right). The ghost bat (D) is an echolocating mammal that relies heavily on its auditory system. A large proportion of its neocortex is devoted to the auditory system (D, left), and the relative size of A1 is large compared to other primary fields (D, right). Scale bar = 1 mm. Squirrel: Kaas et al (1989), Krubitzer et al (1986), Luethke et al (1988). Mouse: Carvell & Simons (1986), Stiebler et al (1997), Wagor et al (1980), Woolsey (1967). Platypus: Krubitzer et al (1995). Ghost bat: Krubitzer (1995), Wise et al (1986).

on its vibrissae than its visual system, much of the neocortex is devoted to the somatosensory system, and in particular, to processing input from the vibrissae. The relative size of S1 is larger in the mouse than the size of either V1 or A1 (Fig. 2B, based on Carvell & Simons 1986, Stiebler et al 1997, Wagor et al 1980, Woolsey 1967). Similarly, the duck-billed platypus has an extremely well developed bill that is composed of densely packed mechanosensory and electrosensory receptors. The platypus uses its bill for most activities including navigating in water, prey capture, predator avoidance and mating. Most of the neocortex in the platypus is devoted to the somatosensory system, and the relative size of S1 compared to other primary fields is quite large; in fact, approximately two-thirds of the cortex is involved in processing input from the bill (Fig. 2C, based on Krubitzer et al 1995). Finally, the ghost bat is an echolocating mammal that relies on its auditory system for most vital behaviours. It is not surprising that a large proportion of its neocortex is devoted to the auditory system, and that the relative size of A1 is large compared to other primary fields (Fig. 2D, based on Krubitzer 1995, Wise et al 1986). In addition to these, other mammalian species show the same trend, namely, that the relative size of a sensory domain and the primary cortical area within that domain are related to the behavioural relevance of that sensory system. These and other types of modifications, such as the addition of cortical fields, changes in connectivity and the addition of modules, constitute a limited number of the systems level changes that have occurred in evolution. Presumably, these modifications account for the high degree of variability in sensory processing and related behaviours observed across mammals.

The comparative approach has yielded important insights into cortical field evolution. In particular, it has allowed us to identify a homologous constellation of cortical fields and their connections, and to appreciate the types of modifications that have been made to the brain throughout evolution. Furthermore, it has revealed that the number of modifications made to the neocortex appears to be constrained and that some of these modifications, such as cortical domain allocation and cortical field size, appear to be related to specialized morphology and use.

How is the developmental approach utilized to study cortical evolution?

A developmental approach can be used to determine how cortical domains and cortical field size have changed in relation to peripheral, morphological specializations and use. This approach can also be used to uncover the mechanisms that give rise to aspects of cortical field organization, as well as to understand how these mechanisms have been altered in different lineages to account for phenotypic variability. In general, studies of the developing nervous system that seek to understand how structures and areas emerge and how these areas become precisely interconnected fall into two main categories.

The first category includes studies that examine the intrinsic or genetic contribution to aspects of neocortical development. This group is varied and includes descriptions of spatial and temporal aspects of the normal developmental processes that are thought to be intrinsic to the neocortex (Donoghue & Rakic 1999, Rubenstein et al 1999), such as the assignment of the rostrocaudal axis (Fukuchi-Shimogori & Grove 2003, Muzio & Mallamaci 2003), the formation of thalamocortical connections (Bishop et al 2000, Inoue et al 1998) and the emergence of particular architectonic features (Fukuchi-Shimogori & Grove 2001, Hamasaki et al 2004). Most of these types of studies alter the genetic environment of a developing animal via mutations, over-expression, or ectopic placement of a gene or gene product. One problem with these types of studies is that they are often confounded, since a single gene or gene product is usually involved in several different processes at multiple stages of development. Furthermore, while genetic mutations are an integral part of evolution, many of the mutations that are studied in developmental experiments result in offspring that do not survive postnatally. Since these types of mutations would result in non-viable offspring, the evolutionary relevance of some genetic models is unclear. However, an extremely important strength of this approach is that it allows us to directly examine potential genetic mechanisms that give rise to cortical attributes in development and evolution. Another method used to examine the types of intrinsic changes that give rise to phenotypic variability is to physically alter some aspect of the developing neocortex (e.g. Huffman et al 1999, Schlaggar & O'Leary 1991) and determine whether the resulting changes that are observed in cortical organization are consistent with the types of changes that occur in evolution. The advantage of this approach is that an area of interest can be manipulated directly, without the confounds associated with global genetic changes. The goal of physical manipulations is to understand whether specific changes that are made in the developing cortex can induce the formation of cortical phenotypes that are similar to those observed in extant lineages. This approach does not test the mechanism that may naturally cause these types of manipulations; it only looks at the results of the manipulations. This is different from the genetic manipulation approach described above, which examines genetic mechanisms that cause the variation observed in extant lineages. Nonetheless, both approaches are aimed at understanding the intrinsic or genetic contribution to neocortical development.

The second category includes studies that examine the role of extrinsic factors in generating aspects of neocortical organization, such as sensory-driven activity. There are two main methods used to study the extrinsic contributions to the cortical phenotype. First, physical manipulations can be made to the developing sensory receptor array, such as enucleating an eye or removing vibrissae. As described above, the advantage of physical manipulations is that they can be made directly to specific structures, without the confounds of changing a global developmental factor.

Another method that can be used to examine the role of extrinsic factors in generating cortical phenotype is to change the external environment in which the animal develops. The advantage to this approach is that it does not cause any genetic or physical confounds with the animal's normal developmental process. Furthermore, this approach mimics natural processes and can provide some insights into how alterations in the nervous system and behaviour are shaped by the environment and, ultimately, incorporated into the genome via natural selection (see Krubitzer & Kaas 2005). Unfortunately, it is difficult, although not impossible, to generate experiments that can fully explain developmental mechanisms by only manipulating the environment.

Because all of the techniques that are used to examine the intrinsic and extrinsic contributions to the phenotype have limitations associated with them, it is best to use a combination of these approaches to understand how structures and areas emerge in the developing nervous system and how those areas evolve. Together, these types of studies have already uncovered a number of important developmental mechanisms that may be responsible for the types of modifications that have been made to the neocortex, such as cortical domain allocation, cortical field size determination and connectivity.

Intrinsic mechanisms that shape cortical field development and evolution

One way that aspects of neocortical organization, including geographic location, patterns of connections, relative cortical field size and module formation can be changed is by altering genes intrinsic to the neocortex. For example, transcription factors, such as *Emx2* and *Pax6,* appear to play an important role in assigning the geographic relationships between primary fields in the rostrocaudal axis and the patterning of thalamocortical connections (Bishop et al 2000, Muzio & Mallamaci 2003, see O'Leary & Nakagawa 2002). In mice lacking *Emx2*, thalamic afferents from the ventral posterior nucleus (VP), which normally innervate S1, are shifted far caudally, into cortex that would normally develop into visual cortex (Fig. 3A, Bishop et al 2000), demonstrating that changes in gene expression can play an important role in the patterning of thalamocortical connections. In terms of cortical field size and location, mice genetically engineered to overproduce nestin-*Emx2* have a larger V1 than wild-type animals and other primary fields, such as S1, shift rostrally on the cortical sheet (Fig. 3C, Hamasaki et al 2004). Finally, when the signalling protein FGF8 (fibroblast growth factor 8), which is involved in setting up anterior–posterior patterning via the regulation of *Emx2* expression (Fukuchi-Shimogori & Grove 2003) and is normally located in the rostral pole of the neocortex, is electroporated into an ectopic location caudal to S1, a duplicate cortical barrel field (defined histochemically) is observed just caudal to S1 (Fig. 3E, Fukuchi-Shimogori & Grove 2001). This suggests that altering patterns of gene expression

Genes intrinsic to the Neocortex   **Peripheral Morphology/Activity**

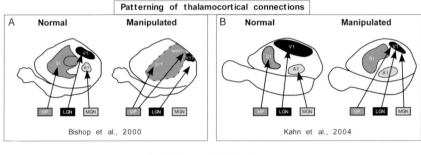

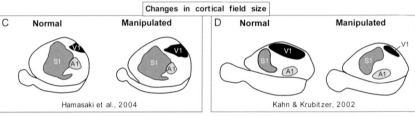

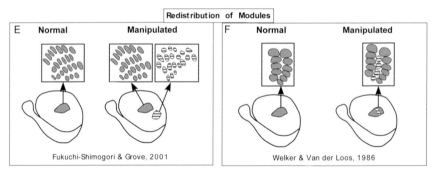

FIG. 3.   Similar types of system level changes can be induced in cortical development through intrinsic genetic mechanisms or through extrinsic changes in peripheral morphology and activity. First, thalamocortical connections can be shifted caudally in transgenic mice by altering the expression of the *Emx2* gene (A). Alternatively, thalmocortical afferents can be shifted by changing peripheral morphology, such as bilaterally enucleating the eyes early in development (B). Another consistent modification made to the neocortex in different lineages has been a change in the size of a cortical field. The size of a cortical field can be changed by overexpressing genes, such as the overproduction of nestin-*Emx2* in transgenic mice that results in a larger V1 than wild-type animals (C). Alternatively, cortical field size can be decreased by altering peripheral morphology by bilaterally enucleating the eyes early in development of opossums (D). Although the direction of change is different, in the former study V1 increases and in the later it decreases, both types of manipulations result in changes in the size of V1. Finally, another type of modification that can be made to the neocortex is the addition of modules. This has been accomplished experimentally by electroporating the signalling molecule FGF8 into an ectopic location in cortex, caudal to the normal location of the barrel fields (E) or by selectively breeding mice to grow an extra row of whiskers (F). In both experiments, additional barrels were generated.

can restructure the formation of modules on the cortical sheet. Taken together, these studies in developing animals indicate that several of the ubiquitous features of cortical organization, such as thalamocortical and corticocortical connectivity, primary sensory field size and location, and modular organization, can be genetically regulated.

Extrinsic mechanisms that shape cortical field development and evolution

Alternatively, altering peripheral morphology and sensory driven activity early in development can also have a dramatic affect on neocortical organization (Kahn et al 2004, Rakic et al 1991, Sur & Leamey 2001 for review). Alterations in peripheral morphology can influence domain allocation, the size of a cortical field, and the development of thalamocortical afferents. For example, animals that were bilaterally enucleated early in development, prior to thalamocortical innervation, had massive changes in cortical organization and architecture, in that there was a large change in sensory domain allocation. All of what would be visual cortex, including V1 was taken over by the somatosensory and auditory systems. Furthermore, the size of V1 was greatly reduced in enucleated animals (Fig. 3D, Kahn & Krubitzer 2002, Rakic et al 1991), and alterations in thalamocortical connections were observed (Kahn et al 2004). In experimental animals, primary visual area (V1), as defined by electrophysiology and myeloarchitecture and which is normally connected only with the lateral geniculate nucleus (LGN), formed connections with the LGN, the ventral posterior nucleus (VP), and the medial geniculate nucleus (MGN) (Fig. 3B). Interestingly, in this study in *Monodelphis domestica*, as well as in a previous study by Rakic et al (1991), it was observed that bilateral enucleation early in development leads to the emergence of a new architectonic area, called area X, that appears between areas 17 and 18. Area X has been shown to process auditory and somatic inputs, primarily from the head, vibrissae and snout (Kahn & Krubitzer 2002). These results suggest that early changes in peripheral morphology, such as bilateral enucleation, can have an enormous effect at multiple levels of the nervous system. While the fact that V1 is still present in enucleated animals indicates that some aspects of arealization are intrinsically regulated and do not depend on specific sensory inputs, alterations in domain allocation, cortical field size, connectivity and the emergence of a new cortical area indicate that extrinsic mechanisms play an equally important role in cortical field development and evolution.

In addition to determining the connections and relative size of primary cortical areas, changes in peripheral morphology can also influence modular organization. Early work by Welker & Van der Loos (1986) demonstrated that mice that were selectively bred to have an extra row of whiskers developed an additional row of barrels that were inserted within the barrel field of S1 (Fig. 3F, see also Catania & Kaas 1997). Finally, alterations in the type and amount of sensory stimulation that the developing nervous system is exposed to can have a dramatic affect on the resulting phenotype. For example, rats reared in a chronically noisy environment had a large disruption in the development of A1, and those reared with exposure to pure tones developed frequency maps in A1 that showed a specific expansion of the pure tone frequency (Chang & Merzenich 2003, Zhang et al 2002, 2003). Taken together, these studies indicate that several features of cortical organization, sensory domain allocation, cortical field size, thalamocortical connectivity and modular organization can be regulated by peripheral morphology and sensory-derived activity.

Conclusions

Since the same types of modifications (e.g. changes in cortical field size) that occur naturally in the neocortex can be accomplished in more than one way, it is critical to examine the roles of both intrinsic genetic factors and extrinsic factors that contribute to different attributes of the developing nervous system, and ultimately, that generate phenotypic diversity in different lineages. For example, we can test the importance of intrinsic genetic factors by comparing patterns of gene expression in highly derived animals, such as an echolocating bat and a mouse. These animals have a similar size neocortex but different, highly derived peripheral morphologies which are specialized for processing different modalities of sensory information. Furthermore, the neocortex of these animals is markedly different in terms of sensory domain allocation and the relative size of primary fields (Fig. 2). Given that all mammals possess a common plan of organization, it seems reasonable to conclude that the presence of the primary cortical areas, their connections, their modular organization and their size are, in large part, genetically determined. The caveat to this is that the relationship between specific patterns of gene expression and cortical field development is still unclear. Nevertheless, if genes play a significant role in patterning the cortex, then we would expect animals with similar sized cortices and noticeably different sensory domains, such an echolocating bat (Fig. 2D) and a highly somatic animal like the mouse (Fig. 2B), to have the same genes regulating general features of cortical development, yet to have slightly different gradients of expression or, possibly, differences in temporal expression. In other words, the same genes most likely control cortical patterning in all mammals; the differences between species are probably derived from slight differences in how and when those genes are expressed. Therefore, by comparing the patterns of gene expression in highly derived animals, we can deduce which genes are actually involved in assigning cortical domains and understand how and when the expression patterns of those genes are altered during development in different lineages.

In conclusion, by combining the comparative approach with a developmental approach it is possible for us to study the roles of genes and sensory-driven activity derived from peripheral morphology on the formation of cortical fields in development, and to make accurate inferences about how these mechanisms contribute to neocortical evolution. Not surprisingly, changes in intrinsic patterns of gene expression or in the sensory receptor arrays in the periphery can have remarkably similar effects on cortical field formation, indicating that the same types of phenotypic modifications can be accomplished via different mechanisms. Only by using a multidisciplinary approach can we hope to unravel the mechanisms used in evolution, in different lineages, and to understand how these mechanisms are altered in nature to generate the wide range of diversity seen in mammalian species.

Acknowledgements

We thank Deborah Hunt for her helpful comments on this manuscript. This work was supported by a McDonnell grant and an NINDS award (R01-NS35103) to Leah Krubitzer and an NSF fellowship (DG-0202740) to Sarah Karlen.

References

Bishop KM, Goudreau G, O'Leary DD 2000 Regulation of area identity in the mammalian neocortex by Emx2 and Pax6. Science 288:344–349

Bronchti G, Heil P, Sadka R, Hess A, Scheich H, Wollberg Z 2002 Auditory activation of "visual" cortical areas in the blind mole rat (Spalax ehrenbergi). Eur J Neurosci 16:311–329

Bullock TH 1984 Comparative neuroscience holds promise for quiet revolutions. Science 225:473–478

Campos GB, Welker WI 1976 Comparisons between brains of a large and a small hystricomorph rodent: capybara, Hydrochoerus and guinea pig, Cavia; neocortical projection regions and measurements of brain subdivisions. Brain Behav Evol 13:243–266

Carvell GE, Simons DJ 1986 Somatotopic organization of the second somatosensory area (SII) in the cerebral cortex of the mouse. Somatosens Res 3:213–237

Catania KC 2000 Cortical organization in insectivora: the parallel evolution of the sensory periphery and the brain. Brain Behav Evol 55:311–321

Catania KC 2002 Barrels, stripes, and fingerprints in the brain—implications for theories of cortical organization. J Neurocytol 31:347–358

Catania KC, Kaas JH 1997 The mole nose instructs the brain. Somatosens Mot Res 14:56–58

Chang EF, Merzenich MM 2003 Environmental noise retards auditory cortical development. Science 300:498–502

Cooper HM, Herbin M, Nevo E 1993 Visual system of a naturally microphthalmic mammal: the blind mole rat, Spalax ehrenbergi. J Comp Neurol 328:313–350

Donoghue MJ, Rakic P 1999 Molecular evidence for the early specification of presumptive functional domains in the embryonic primate cerebral cortex. J Neurosci 19:5967–5979

Ehret G 1997 The auditory cortex. J Comp Physiol [A] 181:547–557

Fukuchi-Shimogori T, Grove EA 2001 Neocortex patterning by the secreted signaling molecule FGF8. Science 294:1071–1074

Fukuchi-Shimogori T, Grove EA 2003 Emx2 patterns the neocortex by regulating FGF positional signaling. Nat Neurosci 6:825–831

Hamasaki T, Leingartner A, Ringstedt T, O'Leary DD 2004 EMX2 regulates sizes and positioning of the primary sensory and motor areas in neocortex by direct specification of cortical progenitors. Neuron 43:359–372

Heil P, Bronchti G, Wollberg Z, Scheich H 1991 Invasion of visual cortex by the auditory system in the naturally blind mole rat. Neuroreport 2:735–738

Henry EC, Marasco PD, Catania KC 2005 Plasticity of the cortical dentition representation after tooth extraction in naked mole-rats. J Comp Neurol 485:64–74

Huffman KJ, Molnar Z, Van Dellen A, Kahn DM, Blakemore C, Krubitzer L 1999 Formation of cortical fields on a reduced cortical sheet. J Neurosci 19:9939–9952

Inoue T, Tanaka T, Suzuki SC, Takeichi M 1998 Cadherin-6 in the developing mouse brain: expression along restricted connection systems and synaptic localization suggest a potential role in neuronal circuitry. Dev Dyn 211:338–351

Jerison HJ 1973 Evolution of the brain and intelligence. Academic Press, New York xiv p 482

Johnson JI 1990 Comparative development of somatic sensory cortex In: Jones EG, Peters A, editors Cerebral Cortex. Plenum, New York, p 335–449

Johnson JI, Kirsch JA, Reep RL, Switzer RC, 3rd 1994 Phylogeny through brain traits: more characters for the analysis of mammalian evolution. Brain Behav Evol 43:319–347

Kaas JH 1982 The segregation of function in the nervous system: Why do the sensory systems have so many subdivisions? Contrib Sens Physiol 7:201–240

Kaas JH 1983 What, if anything, is SI? Organization of first somatosensory area of cortex. Physiol Rev 63:206–231

Kaas JH 2005 From mice to men: the evolution of the large, complex human brain. J Biosci 30:155–165

Kaas JH, Krubitzer LA, Johanson KL 1989 Cortical connections of areas 17 (V–I) and 18 (V–II) of squirrels. J Comp Neurol 281:426–446

Kahn DM, Krubitzer L 2002 Massive cross-modal cortical plasticity and the emergence of a new cortical area in developmentally blind mammals. Proc Natl Acad Sci USA 99:11429–11434

Kahn DM, Long SJ, Krubitzer L 2004 Aberrant cortical connections in developmentally blind mammals (Monodelphis domestica). Society for Neuroscience Program No 83916

Krubitzer L 1995 The organization of neocortex in mammals: are species differences really so different? Trends Neurosci 18:408–417

Krubitzer L 2002 Evolutionary perspectives in: cognitive neuroscience. In: Gazzaniga MS, Ivry RB, Mangun GR (eds) The biology of the mind. Norton, New York, p 577–596

Krubitzer L, Huffman KJ 2000 Arealization of the neocortex in mammals: genetic and epigenetic contributions to the phenotype. Brain Behav Evol 55:322–335

Krubitzer L, Kaas J 2005 The evolution of the neocortex in mammals: how is phenotypic diversity generated? Curr Opin Neurobiol 15:444–453

Krubitzer L, Kahn DM 2003 Nature versus nurture revisited: an old idea with a new twist. Prog Neurobiol 70:33–52

Krubitzer L, Manger P, Pettigrew J, Calford M 1995 Organization of somatosensory cortex in monotremes: in search of the prototypical plan. J Comp Neurol 351:261–306

Krubitzer LA, Sesma MA, Kaas JH 1986 Microelectrode maps, myeloarchitecture, and cortical connections of three somatotopically organized representations of the body surface in the parietal cortex of squirrels. J Comp Neurol 250:403–430

Levitt P, Eagleson KL 2000 Regionalization of the cerebral cortex: developmental mechanisms and models. Wiley, Chichester (Novartis Found Symp 228) p 173–181; discussion 181–177

Luethke LE, Krubitzer LA, Kaas JH 1988 Cortical connections of electrophysiologically and architectonically defined subdivisions of auditory cortex in squirrels. J Comp Neurol 268:181–203

Murphy WJ, Pevzner PA, O'Brien SJ 2004 Mammalian phylogenomics comes of age. Trends Genet 20:631–639

Muzio L, Mallamaci A 2003 Emx1, emx2 and pax6 in specification, regionalization and arealization of the cerebral cortex. Cereb Cortex 13:641–647

O'Leary DD, Nakagawa Y 2002 Patterning centers, regulatory genes and extrinsic mechanisms controlling arealization of the neocortex. Curr Opin Neurobiol 12:14–25

Rakic P, Suner I, Williams RW 1991 A novel cytoarchitectonic area induced experimentally within the primate visual cortex. Proc Natl Acad Sci USA 88:2083–2087

Reep RL, Johnson JI, Switzer RC, Welker WI 1989 Manatee cerebral cortex: cytoarchitecture of the frontal region in Trichechus manatus latirostris. Brain Behav Evol 34:365–386

Rosa MG, Krubitzer LA 1999 The evolution of visual cortex: where is V2? Trends Neurosci 22:242–248

Rubenstein JL, Anderson S, Shi L, Miyashita-Lin E, Bulfone A, Hevner R 1999 Genetic control of cortical regionalization and connectivity. Cereb Cortex 9:524–532

Schlaggar BL, O'Leary DD 1991 Potential of visual cortex to develop an array of functional units unique to somatosensory cortex. Science 252:1556–1560

Stiebler I, Neulist R, Fichtel I, Ehret G 1997 The auditory cortex of the house mouse: left-right differences, tonotopic organization and quantitative analysis of frequency representation. J Comp Physiol [A] 181:559–571

Sur M, Leamey CA 2001 Development and plasticity of cortical areas and networks. Nat Rev Neurosci 2:251–262

Van der Loos H, Welker E, Dorfl J, Rumo G 1986 Selective breeding for variations in patterns of mystacial vibrissae of mice Bilaterally symmetrical strains derived from ICR stock. J Hered 77:66–82

Wagor E, Mangini NJ, Pearlman AL 1980 Retinotopic organization of striate and extrastriate visual cortex in the mouse. J Comp Neurol 193:187–202

Wise LZ, Pettigrew JD, Calford MB 1986 Somatosensory cortical representation in the Australian ghost bat, Macroderma gigas. J Comp Neurol 248:257–262

Woolsey TA 1967 Somatosensory, auditory and visual cortical areas of the mouse. Johns Hopkins Med J 121:91–112

Zhang LI, Bao S, Merzenich MM 2002 Disruption of primary auditory cortex by synchronous auditory inputs during a critical period. Proc Natl Acad Sci USA 99:2309–2314

Zhang LI, Tan AY, Schreiner CE, Merzenich MM 2003 Topography and synaptic shaping of direction selectivity in primary auditory cortex. Nature 424:201–205

DISCUSSION

Diamond: There are far more species now extinct than those currently existing. We can assume that extinct species didn't survive because something didn't work as efficiently as it needed to work. Presumably, cortical evolution is interesting because what didn't work in extinct species might in many cases have been something about brain organization. Could you guess about some of the cortical organizations that extinct species might have had, but which didn't work?

Krubitzer: It would be dependent on the context in which that species was evolving. My guess is that it wasn't an experiment in brain organization but it had more to do with a rapid change in the environment. What I hope I have demonstrated is that phenotypic variability is in part genetically driven and in part activity dependent. The intracellular mechanisms that allow for plasticity are probably genetically determined, but the phenotypes that unravel aren't, which means they masquerade as evolution. As long as the environment in which that individual is unravelling is constant, that phenotype is going to look the same. An important property of the mammalian neocortex is that it is plastic. If there are small changes in the environment, the cortex is going to be able to adapt fairly well. One thing we are now doing is looking at the behaviour of these animals, to see what an expanded visual cortex is doing in congenitally deaf mice. In terms of the extinctions, it is not so much that they have mucked around with changing cortical organization, but the environment has changed dramatically. Even if there is some normal distribution of phenotypes that could be generated under some normal distribution of environmental conditions, if the environmental fluctuations become extreme and fall outside of the distribution of potential changes that could be made to neural and non-neural tissue, then extinction occurs. When I am talking about activity or external influences on phenotype, I am talking about sensory inputs and so on, but even passive changes such as in pH, body temperature or amount of food present can factor into extinction.

Logothetis: Can you replay your theory in slow motion? I missed most of it because you were talking so fast. You said that we have some kind of organization and then there is some kind of input coming in. In the beginning it is sparse and the input is reaching existing areas, and these are cells among other cells. Is this what you are showing?

Krubitzer: No. What I'm showing is there is a given pattern of thalamic activity, from the ventral posterior nucleus or LGN, for example, imposed on the cortical sheet in a particular animal. Let's say that within some lineage some change occurs in peripheral morphology. For example, I develop a highly sensitive vibrissal system. This change promotes changes in thalamic organization, and in turn, I develop a modular organization or some segregation of those inputs in the neocortex. It may be beneficial to keep those inputs separate (e.g. it helps me make finer discriminations of roughness and so on). This sort of change is constantly occurring. Perhaps I have an evolution of a new receptor type, for instance a new ganglion cell type. This will have an influence of the next structure, in this case the LGN, which in turn is going to change cortical organization. For the somatosensory system, a very small change in a small piece of tissue such as the dorsal column nucleus is going to have a large cortical signature. I don't think you need big tweaks in peripheral morphology to generate these changes in the cortex.

Logothetis: I am still not clear about what your theory is. You are showing small targets interspersed in the cortex, and then at some point these may aggregate in the cortex on the basis of similarity.

Krubitzer: Let me try to illustrate this. If we look at something like V2, people have argued that it is not a single representation but is actually three representations interdigitated. Each of the bands in V2 obey all the rules of a cortical field. In some species the segregation may continue because it works out.

Diamond: Maybe it will help if you define what an alternative theory might be.

Krubitzer: An alternative theory might be that there is something inherent about V1: it is absolutely genetically specified and there is some change in spatial expression of genes or the temporal pattern of gene expression that leads to a new cortical field.

Logothetis: What you have just proposed as an alternative doesn't contradict your theory.

Krubitzer: My theory assumes that cortical fields are not generated simply by mechanisms intrinsic to the cortex. Instead, connections and activity are involved in generating new cortical fields or new patterns of organization. If you take a strictly genetic approach, you would have to argue that some change in the spatial and/or temporal pattern of expression of some gene, or the addition or deletion of an allele is solely responsible for some aspect of cortical organization and the emergence of cortical fields in development.

Logothetis: You are saying that the genetic mechanisms determine the range.

Krubitzer: Yes. I was pointing out that there are large genetic constraints which means that you can't eliminate particular fields, such as V1, their location (in caudomedial cortex) is for the most part invariant, and aspects of thalamocortical and corticocortical connectivity are similar in all mammals.

Logothetis: You seem to be suggesting that at some point the cells decide they have a greater advantage if they are closer to each other and share connectivity than if they are dispersed.

Krubitzer: Exactly. At some point I would rather be mapped next to the representation of koniocellular cells, than next to the representation of M cells that share a similar portion of retinal space. It is a compromise that depends on what will work for that particular animal. All of this is inference.

Derdikman: Can we gain insights about sensorimotor function from looking at the evolution of the sensorimotor areas and comparing them to behaviour within species?

Krubitzer: Yes. Although some fields may be homologous, such as V1 and M1, it doesn't mean that they are analogous (i.e. have the same function). Most people at this meeting have been discussing single units and looking at behaviour. They then correlate activity with some aspect of behaviour or type of sensory stimulation. That unit sits within a cortical field which sits within a network. In the platypus, there is a small V1, no apparent V2, a small undifferentiated LGN, and a micro ophthalmic eye. In monkeys, V1 is present as well as V2 and several other extrastriate areas to which V1 projects. The LGN is laminated and highly differentiated, and the eye is highly developed. V1 projects to and receives feedback from several extrastriate cortical areas; this is not the case for the platypus. Is V1 doing the same thing in a platypus as it is in a monkey? I doubt it.

Treves: I am trying to think of something that is *not* in your theory. You showed us a lot of evidence about cortical flexibility or adaptation, and some nice evidence about thalamic flexibility. Can you contrast this cortico-thalamic readiness for change with subcortical structures that are more genetically hardwired?

Krubitzer: There aren't a lot of data on this. If you look across a lot of mammals, the cortex has changed most dramatically. The thalamus has changed quite a bit as well. But brainstem structures haven't changed as much, because they are doing things that are important for survival such as breathing and regulating heartbeat. Those genes involved in specification for parts of the brain that are necessary for life function, are also necessary for non-life sustaining functions, such as aspects of cortical arealization. This leads to serious functional integration, and places huge constraints on changing particular aspects of the non-life sustaining features of organization (i.e. cortical organization) that this gene encodes. Thus, the brain is a compromise.

Brecht: In your theory there are modules that are then sorted out to areas. This is not what we feel when we look at the rodent work. There the modules appear to be something very different from the maps. You can have modules or barrels, or you don't have them, but you always have the map. More importantly you can genetically make a map. You put on FGF8 and then you have second barrel cortex: this is a genetic mechanism that is being read by the afferent axons. Do you know of any instance where there is evidence for modules that then get sorted out into two maps?

Krubitzer: No, but it would be difficult to actually observe such a thing given the large time scales of evolution. The point I was making earlier about the barrel business and whisking versus not whisking is that the barrels are in some sense epiphenomenal. They are not needed to do certain things. Jonathan Horton recently showed in squirrel monkeys half of them had ocular dominance columns (OCDs) and half of them didn't, which suggests that OCDs are not necessary for particular aspects of visual processing.

Brecht: Doesn't the whole developmental work suggest maps are something very different from modules?

Krubitzer: You can de-correlate modules with function and you can also de-correlate architectonically defined cortical areas such as S1 or V1 with function. In most mammals, primary cortical field function and architecture are highly correlated.

Haggard: I'd like to encourage you to speculate. You have been talking mainly about primary cortical areas. Particularly in the context of decision making, what will these tell us about secondary cortical areas? Traditionally these have been thought of as more abstract, and didn't seem to have this close allegiance to the periphery.

Krubitzer: That's a good point. I talked almost exclusively about primary areas. Developmental neurobiologists do the same because they are really easy to identify. My gut tells me that there is something different about primary fields versus secondary cortical areas. Primary fields may be more genetically regulated or constrained than other fields. If we look at most non-primary cortical areas architec-tonically, they aren't particularly distinct. I think these fields are more susceptible to environmental influences during development than in adults. It is not a coincidence that as we examine extrastriate cortex in macaque monkeys, everyone argues about what the boundaries of cortical fields are. This may be because in the experiments in which we are trying to figure out what particular cortical fields are doing, and how cortex is subdivided, by training the animal to perform some experimenter driven task, we are actually changing the cortex. In developing animals I think this cortex is extremely plastic. We did an experiment many years ago where we mapped somatosensory cortex in a one day old monkey. We found that S1 was in place, but we couldn't find area 1 or area 2. This indicates to me that the environment (i.e. patterns of sensory receptor array activity) play a larger role in directing the organ-ization of these fields than it does for primary fields.

Haggard: You have described a mechanism for the module to break away, aggre-gate and make a new area. Would this allow secondary areas to arise from primary ones? Wouldn't it just give you more primary ones. The evolution of secondary areas might need a different mechanism.

Krubitzer: Part of me thinks that primary fields such as S1 and V1 are the 'off-ramp' for change, in that all new fields may originate in primary fields and segre-gate or move out of them. I agree that the mechanisms I describe for primary cortex may not be the same for association cortex.

Sparks: You talked about anatomical controls to show that the inputs are the same, but I didn't hear you talking about anatomical experiments to show that the outputs of areas were the same. This relates to a general concern I have about 'remapping' and adaptive remapping. Is it really adaptive? As the cortical area is invaded and for-merly visual cells are now activated by auditory stimuli, is the sensation evoked still vision or is it auditory? Is the attribution of the activity changed or is only the stim-ulus that induces the activity changed? By looking at output mappings you could get a handle on this.

Krubitzer: That's a good idea. One other thing we have been doing is looking at behaviour in a more global sense: is it adaptive? Are these animals better at per-forming some tasks with this expanded cortex? One of the things we have also started to do is to map the reorganized areas more closely. For example, in the bilat-erally enucleated animals almost all of reorganized cortex contains representations of the vibrissae of the head and face, which overlaps with auditory inputs. To me this suggests that there is some functional dependence between the vibrissae of the face and the auditory system for localization that doesn't exist in normal animals. Your question will be difficult to answer even if we look at outputs. The best way we can get at this is to look at humans who have lost a sensory system. Unfortu-nately I don't think their cortex gets invaded to the same extent unless their loss is congenital.

General discussion III

Tanifuji: In relation to decision making, I found Giacomo Rizzolatti's paper very interesting. Yesterday and today we have been talking about decision making as a response to sensory input. You have shown that monkeys can interpret the actions of others, and this is also information that can induce action. If this is the case, it is likely to have some bias to particular actions. For example, when someone picks up food, one's behaviour may make me feel hungry but does not necessarily elicit my action to pick up food. But there are some actions that really evoke my action. If someone attacks me, then I would escape from them. In this case the perceived action and action of execution are not the same, but the perceived action and the executed action have strong behavioural or social relevance. Similarly, couldn't you suggest some behavioural meaning as to why certain actions are represented in mirror neurons?

Rizzolatti: There is an interesting ethological study carried out by Visalberghi on this issue (Ferrari et al 2005). She showed that if a monkey (a Cebus) sees another monkey eating, it starts to eat even if already satiated. The decision to eat is triggered by seeing another monkey eating.

Tsakiris: How would you expect the mirror neuron system to behave in the case where I have to observe an action but at the same time perform an opposite action?

Rizzolatti: There is an experiment by the Munich group relevant to this (Brass et al 2000). They showed that when individuals are instructed to do the same action, for example lifting a finger, as the one they are observing, they respond fast. On the contrary, when they have to do the opposite action, they are slow and prone to mistakes. This implies a common perceptual-motor space.

Tsakiris: In that case you would say that the mirror neuron system represents mostly the actions of someone else.

Rizzolatti: Actions of others activate the mirror system of the observer. In this way the observer understands the seen action. Subsequently, he/she makes a decision on what to do.

Romo: Many studies have shown that action can be initiated in the absence of an external signal. How does your system relate to this kind of work? In your case, imitation of an action is related to perception, but in other conditions you can generate an action by simply imagining an action.

Rizzolatti: It was very popular some years ago to distinguish sharply between self-generated and stimulus-triggered responses. In the category of stimulus-triggered responses were included not only (quite correctly) reflexes, but also voluntary actions such as grasping an object. In my opinion this distinction is wrong. Actions on objects are self-generated exactly as actions in which no object is present (e.g. lifting the arm). In both cases the acting individual has to decide first that he wants to act, to retrieve then the desired action from the action vocabularies located in the parieto-frontal circuits and to allow, finally, that the selected action be implemented by motor areas and subcortical centres. In both cases the 'braking system' formed by mesial areas (pre-SMA and cingulate areas), and controlled by the frontal lobe, is crucial in allowing action execution. The only difference is that, in the case of object-related actions, the action vocabulary is addressed by external stimuli, in the case of action without object it is addressed endogenously. This renders it more difficult to act without objects. The presence of an external stimulus may be a 'convincing' argument in favour of the decision to act. The decisional mechanism, however, remains the same in both cases.

Harris: You mentioned monkeys seeing other monkeys eating and then deciding to start eating themselves. Do you know that it is a decision-based response? For example, we yawn when we see someone else yawn: we are not making decisions to do this.

Rizzolatti: This is a very interesting question. There are indeed behaviours, called contagious behaviours, in which individuals are almost compelled to repeat the observed actions. Yawning is an excellent example of this type of behaviour. I think that in the case of monkeys starting to eat when they see other monkeys eating, there is a clear 'contagious' component, but I am not sure that this fully explains their behaviour. My opinion is that there is also a voluntary component.

Logothetis: What does everyone here think is the minimum condition for an area or a cell to be characterized as correlating to decision making? What behaviour do you have in mind? Should it be working on the basis of thresholds or probability summation, or coincidences?

Diamond: One criterion, albeit one that is hard to demonstrate experimentally, is that manipulating that cell's activity causes a change in decisions. That is, external manipulation of the neural activity causes a change in the decision.

Romo: In my own view to really show the neural computation associated with perceptual decisions, you need to decode the sensory from the working memory processing. If you simply manipulate the sensory input you will not understand the contribution of the working memory component. You need to manipulate both.

Diamond: David Sparks, could you consider the collicular cells that produce eye movements as part of the decision-making process?

Sparks: We placed the animal in a situation in which the duration or probability of the stimulus is manipulated so that the probability of a saccade to the stimulus varied. The collicular cells I was talking about have a low frequency prelude of activity and then a vigorous high frequency burst of activity. In the experiment I did, if you present the target and the animal fails to generate a saccade toward the target, we often observe the prelude to activity but never observe the burst. If the burst of the cell occurred we always got the movement. Jeff Schall is the one who should be responding to this because he is the one who developed the experimental paradigms to look at the decision making. He has experiments in which he can give go and no-go signals.

Schall: I believe the word 'decision' is too vague to apply to the brain. People decide and they need neurons to do so, but mapping decision onto a neuron or an area or even a brain system is absurd: we need to break the decision into its constituent parts. There is a sensory representation and then sometimes it takes time to accumulate evidence, and then you have to generate a response. We are not being clear enough about these separate stages of processing.

Logothetis: I agree we are not being clear enough. It is not absurd to talk about decision making the way that we sometimes do, though. What is needed for the final command to act?

Diamond: I think my comment was degraded at several steps. I thought the question was: What can we take to be evidence that a particular set of neurons is involved in the decision-taking process? You reduced it to an absurdity: which neuron is taking the decision? One of the comments was that it would be trivial to change the activity of visual cortical neurons and make the subject take a different decision. Yet, I would be surprised if you could manipulate externally the activity of visual cortical neurons and actually make animals take a different decision, because I don't think people are able to produce activity in the visual cortex that the animal would use to make a different decision.

Rumiati: Giacomo, do you have any information about whether neurons become mirror neurons at one point in life in the animal? Have you just studied adult animals?

Rizzolatti: Unfortunately, there are no experiments on newborn monkeys. Imitation, however, in babies is already present at birth. This might suggest that there are 'innate' mirror neurons.

Rumiati: How do you think the mirror neuron system relates to memory? Do we use this system to learn things or actions?

Rizzolatti: Mirror neuron system is a memory system for actions. Individual mirror neurons code motor acts. In adult animals many of these neurons are linked one with another forming motor act chains. These chains form a memory system for motor actions. New actions can be learned, by imitation, associating the seen motor acts in new chains. The mirror neuron system does not appear to be involved in learning things. The basic bricks for learning a new action by imitation are the mirror neurons. As I just said, mirror neurons code motor acts. The learning of a new action requires a structure that puts together these bricks in a new way. Our fMRI experiment on learning guitar chords by imitation (Buccino et al 2004) suggests that an important role in the construction of a new action is played by the prefrontal lobe (area 46).

Treves: I have a simple observation. A clear impression I have from putting together the last three papers is that we mammals have simply *not* evolved in order to take decisions in a sophisticated manner. The fact that the beautiful model of Jochen Ditterich can go a long way to describe the experimental results on decision making suggests that the process of decision making itself is not sophisticated enough to explain the richness of mammalian cortical evolution. From Giacomo Rizzolatti's and Leah Krubitzer's papers we have been given the impression that cortical evolution goes towards acquiring a fine representation of the world and the relationship between the components in the environment. Decision making, instead, is something we could do 400 million years ago, before we had a cortex.

Derdikman: To answer Mathew Diamond, I'd suggest that binocular rivalry is an example where the visual percept can influence decision making without the sensory world changing outside.

Schall: In that case what is a decision? With rivalry, you don't choose what to see: it happens to you. There is a sense of decision where if we decide to do something we are responsible for our actions, even legally. Do other animals on the planet decide in the way that we do, deliberating and weighing options? There are lots of choices that we make, where there are alternatives that we decide upon. The brain state in a deliberate decision may be different from that in a choice. With the dots test we heard about earlier, a functional imaging study showed that whereas activation of area MT varied close to linearly with the strength of the motion signal, activation of anterior cingulated cortex occurred only when the motion strength was weak and the alternatives were very ambiguous (Rees et al 2000).

Tanifuji: We may not be able to discuss decision making without taking into account expected outcome or reward for the animal after decision making. In this respect, the case where a certain conflict is involved in the behaviour context could be interesting. For example, if higher reward is expected with lower probability for one choice of behaviour, and lower reward is expected with higher probability for another choice of behaviour, then the interesting question is what would be the animal's decision making and how is neural activity correlated to the monkey's decision making? With a simple task that consists of one stimulus with one action, it may be difficult to discuss this level of question.

But, still I think it is worthwhile to look at the sensory information and how it is used for decision making in such simple tasks. At the early stages, the sensory information representation is designed for efficient analysis of sensory inputs. However in the higher level, representations may be more related to usefulness of the information in decision making. The representation for analysis of sensory stimulus and that for preparation for decision making could be different. One purpose of investigation of sensory information could be to find such difference in representation.

Wolpert: I wonder whether we are making life very hard for ourselves by using these tasks where there is one sensory input leading to one motor output with the decision process in the middle. It is hard to split up where the decision processes are happening. It seems that most of the decisions we make are independent of the senses we get and the actions we are going to make. If I decide I want to get some food, this is a decision I make. There are many routes from the sensory side that lead to this decision and similarly there are many motor acts I do to achieve this. If you can get to areas where decisions are made independent of the actions, we might be able to see the decision process separately from input and output systems.

Gold: The reasons that we link it is because it gives us a chance to know where to put our electrodes. The hope is that we are looking at mechanisms that generalize to the kind of decision making you are talking about.

Dehaene: In response to the challenge of defining what a decision is, it might help to consider when we might apply the term 'decision' to an artificial system. If we look at a thermometer, no one would say that the thermometer decided to track the temperature change. However, I think we would apply the term 'decision' to a computer capable of examining multiple possibilities that are not directly determined beforehand. Decision making is needed only when the possibilities for action are not entirely determined by the sensory information available (as in a reflex). Decision making is a process that mediates between sensory perception and the repertoire of actions, and that may give variable, flexible responses from time to time. Thus, in looking for decision processes in the brain, I would look for whichever processes have flexibility and are sufficiently detached from purely sensorimotor chains of information processing. Decision is a high-level process, very much in the spirit of the Norman-Shallice model which distinguishes between a direct routine pathway and a higher-level system of executive control.

Rizzolatti: Someone raised a question about 'association areas' and their relations with primary sensory areas. Many years ago, Diamond & Hall (1969) proposed that the appearance of association areas preceded that of primary, sensory and motor areas in the evolution of cerebral cortex. I like this idea very much. In the case of the motor system, it is much more logical to admit that the premotor areas, which control the actions globally, appeared first and then, from them, evolved the primary motor cortex, which enables a more fine and subtle control of movements. The same is true for the primary visual cortex. Why do we need small, receptive field in order to have a general idea of the environment? What is the current status of this problem? Which areas did evolve first?

Krubitzer: I think that's a very good point. We all like to think that V1 evolved first and we name our cortical fields accordingly. However, there isn't any evidence that primary fields are the oldest fields, yet this assumption influences how we think the brain works. For example, we think that V1 is the oldest field and other extrastriate cortical areas such as V2, V3 and V4, were added consecutively in evolution and therefore cortical processing is hierarchical. But what if V1 is the newest cortical field? Woolsey named areas S1 and S2, V1 and V2 simply on the basis of when they were discovered. In somatosensory cortex of New World titi monkeys we have identified area 3b (S1) and area 5 (part of posterior parietal cortex). These monkeys also have a very primitive area 1, but they don't have an area 2. Macaque monkeys, who have opposable thumbs, have a very well developed area 1 with small receptive fields, and they have an area 2. So, it appears that an area 2 has been plopped into an existing network which contains the primary field, area 1, and a posterior parietal area.

Harris: A primary field is only defined once we have a secondary field. Essentially, you are talking about things splitting at some point.

Krubitzer: One way to demonstrate if primary fields are evolutionarily older than other cortical fields is to look at the brains of primitive mammals like monotremes, and determine if they only have a V1 and S1. Guess what? They have multiple somatosensory fields and echidnas at least have a V1 and V2. They also have what looks like a small association cortex. So the idea that they only have a V1 and S1 and that primary areas are evolutionarily older has gone by the board.

Porro: But primary areas are larger in primitive mammals, as you showed.

References

Brass M, Bekkering H, Wohlschlager A, Prinz W 2000 Compatibility between observed and executed finger movements: comparing symbolic, spatial, and imitative cues. Brain Cogn 44:124–439

Buccino G, Vogt S, Ritzl A et al 2004 Neural circuits underlying imitation learning of hand actions: an event-related fMRI study. Neuron 42:323–334

Diamond IT, Hall WC 1969 Evolution of neocortex. Science 164:251–262

Ferrari PF, Maiolini C, Addessi E, Fogassi L, Visalberghi E 2005 The observation and hearing of eating actions activates motor programs related to eating in macaque monkeys. Behav Brain Res 161:95–101

Decoding the temporal evolution of a simple perceptual act

Ranulfo Romo, Adrián Hernández, Antonio Zainos, Luis Lemus, Victor de Lafuente, Rogelio Luna and Verónica Nacher

Instituto de Fisiología Celular, Universidad Nacional Autónoma de México, 04510 México, DF, México

Abstract. Most perceptual tasks require sequential steps to be carried out. This must be the case, for example, when subjects discriminate the difference in frequency between two mechanical vibrations applied sequentially to their fingertips. This perceptual task can be understood as a chain of neural operations: encoding the two consecutive stimulus frequencies, maintaining the first stimulus in working memory, comparing the second stimulus to the memory trace left by the first stimulus, and communicating the result of the comparison to the motor apparatus. Where and how in the brain are these cognitive operations executed? We addressed this problem by recording single neurons from several cortical areas while trained monkeys executed the vibrotactile discrimination task. We found that primary somatosensory cortex (S1) drives higher cortical areas where past and current sensory information are combined, such that a comparison of the two evolves into a decision. Consistent with this result, direct activation of the S1 can trigger quantifiable percepts in this task. These findings provide a fairly complete panorama of the neural dynamics that underlies the transformation of sensory information into an action and emphasize the importance of studying multiple cortical areas during the same behavioural task.

2005 Percept, decision, action: bridging the gaps. Wiley, Chichester (Novartis Foundation Symposium 270) p 170–190

Investigations in several sensory systems have shown how neural activity represents the physical parameters of the sensory stimuli both in the periphery and central areas of the nervous system (Hubel & Wiesel 1998, Mountcastle et al 1967, Talbot et al 1968). These investigations have paved the way for new questions that are more closely related to cognitive processing. For example, where and how in the brain do the neuronal responses that encode the sensory stimuli translate into responses that encode a decision (Romo & Salinas 2001, Schall 2001)? What components of the neuronal activity evoked by a sensory stimulus are directly related to perception (Romo et al 1998, Salzman et al 1990)? These questions have been investigated in behavioural tasks where the sensory stimuli are under precise quantitative control and the subjects' psychophysical performances are quantitatively measured (Hernández et al 1997, Newsome et al 1989). One of the main challenges of this approach is that even the simplest cognitive tasks engage a large number of cortical areas, and each one might encode the sensory information in a different way (Romo & Salinas 2003, Romo et al 2004). Also, the sensory information might be combined in these cortical areas with other types of stored signals representing, for example, past experience and future actions (Hernández et al 2002, Romo et al 2002, 2004). Thus, an important issue is to decode from the neuronal activity all these processes that might be related to perception.

Recent studies have provided new insights into this problem using a highly simplified sensory discrimination task (Hernández et al 1997, Mountcastle et al 1990). In particular, these studies have shown the neural codes that are related to sensation, working memory and decision making in this task (Romo & Salinas 2003). An important finding is that primary somatosensory cortex (S1) drives higher cortical areas from the parietal and frontal lobes where past and current sensory information are combined, such that a comparison of the two evolves into a behavioral decision (Romo & Salinas 2003). Another important finding is that quantifiable percepts can be triggered by activating directly the S1 circuit that drives cortical areas associated with perceptual decision making in this task (Romo et al 1998, 2000). Here we discuss the evidence supporting these conjectures.

Psychophysics and neurophysiology of a simple perceptual task

We studied the extracellular activity of single cortical neurons while trained monkeys executed a highly simplified vibrotactile discrimination task (Fig. 1). In this two-alternative, forced-choice task, subjects must compare the frequency of two vibratory stimuli applied sequentially to their fingertips and then use their free hand to push one of two response buttons to indicate which stimulus was of higher or lower frequency. The discrimination task, although apparently simple, is designed so that it can only be executed correctly when a minimum of neuronal operations or cognitive steps is performed: encoding the two stimulus frequencies, maintaining the first stimulus frequency (f_1) in working memory, comparing the second stimulus frequency (f_2) with the memory trace of f_1, and, finally, executing a motor response to report discrimination (Hernández et al 1997). Thus, the discrimination task allows us to investigate a wide range of essential processes of perceptual decision making.

A simple testable hypothesis is that the sequential events associated with the vibrotactile discrimination task are represented in the neuronal activity of a widely distributed system, beginning in S1 and ending in the motor cortices where the motor commands are triggered to report this cognitive operation. It is unlikely that the ascending inputs to S1 encode the essential neuronal computations required to solve this task. Their role could simply be to transmit a neural copy of vibrotactile stimuli, where the stimulus location and features are safely encoded and transmitted

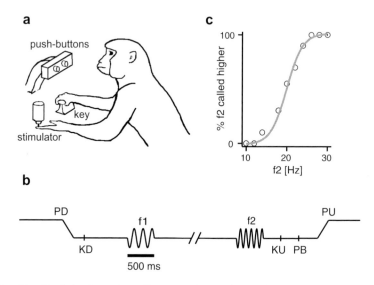

FIG. 1.   The discrimination task. (a) Drawing of a monkey working in the discrimination task.
(b) Sequence of events during discrimination trials. The mechanical probe is lowered, indenting
the fingertip of one digit of the restrained hand; the monkey places its free hand on an immov-
able key (KD); the probe oscillates vertically at the base stimulus frequency (f1); after a delay, a
second mechanical vibration is delivered at the comparison frequency (f2); the monkey releases
the key (KU) and presses either a medial or a lateral push-button (PB) to indicate whether the
comparison frequency was lower or higher than the base. (c) Discrimination perfomance curve,
plotted as the animal's capacity to judge whether f2 is higher than f1.

to S1. Based on these premises, we sought the neuronal activity in several cortical
areas that might be associated with the different components of the vibrotactile
discrimination task. We assumed that in the neuronal responses the stimulus parameters could be quantified and interpreted according to the task demands.

Decoding sensory processes from neuronal activity

In the vibrotactile stimulus range used here (5–50 Hz), mean responses of some S1
neurons (about 30% of the sampled population) typically increase as a monotonic
function of the increasing stimulus frequency (Hernández et al 2000, Salinas et al
2000). For example, during the f_1 period, the firing rate can be approximated to a
linear function: firing rate = $a_1 \times f_1 + b$. Where a_1 and b are constants. The coefficient a_1 is the slope of the rate frequency function, and is a measure of how strongly
a neuron is driven by changes in frequency (in this case f_1). To get an idea of modulation strength, a value of 1 means that the rate increases by 1 spike per second

when frequency increases by 1 Hz. This means that the firing rates of some S1 neurons usually increase with increasing stimulus frequency. Figure 2a shows slope distributions derived from S1 responses. As is illustrated in these plots, this analysis can also be extended to the delay period between the f_1 and f_2 periods. Clearly, none of the S1 neurons that were modulated as a function of the increasing f_1 show a modulation of this type during the delay period. This suggests that S1 neurons do not encode f_1 during the working memory component of the task.

The responses during f_2, where the comparison process takes place, could be an arbitrary linear function of both f_1 and f_2. This could be approximated by the equation: firing rate = $a_1 \times f_1 + a_2 \times f_2 + b$. Fitting this equation to neuronal responses and plotting a_2 as a function of a_1 allows a quantification of the neuron's responses dependent on the f_1 and f_2. Responses that are a function of $f_2 - f_1$ are of particular importance for our ordinal comparison task, since correct responses depend only on the sign of $f_2 - f_1$: $f_2 > f_1$ or $f_2 < f_1$. However, the analysis shows that S1 neurons do not show the comparison process during the f_2 period (Fig. 2a). They increase their firing rate as a function of the increasing f_2, suggesting that the computation between the memory referent of f_1 and the current f_2 input may occur in a central area(s) to S1.

The same analysis used to decode f_1 and f_2 can also be applied to the neuronal responses of areas central to S1. This is an important issue, because it might be possible to quantitatively show sensory information processes in these areas. For example, in secondary somatosensory cortex (S2) similar variations in firing rate are also observed as in S1, however about 40% of the neurons have negative slopes during the f_1 period (Romo et al 2002, 2003, Salinas et al 2000). The firing rates of these neurons decrease as a linear function of the increasing stimulus frequency whereas the remainder have positive slopes and fire more strongly to the increasing stimulus frequency (Fig. 2b). All areas central to S2 that have been examined so far (except for primary motor cortex [M1]) and that are active in the vibrotactile discrimination task show similar monotonic responses and similar proportions of positive and negative slopes (Fig. 2c–f). These areas are the ventral premotor cortex (VPC), prefrontal cortex (PFC) and the medial premotor cortex (MPC) (Brody et al 2003, Hernández et al 2002, Romo et al 2002, 2003, 2004). The responses seem to proceed in a serial fashion, with shorter latencies in S1 than in S2, PFC, VPC and MPC (Romo et al 2004). Whether this reflects a serial or parallel processing is not clear. There is strong evidence that S2 is driven by S1 (Pons et al 1987, Burton et al 1995), but it is not clear whether S2 drives the PFC, VPC and MPC. Some anatomical studies suggest that S2 is connected with these areas, but more studies are needed to establish whether this is so (Cipolloni & Pandya 1999, Disbrow et al 2003, Godschalk et al 1984). Thus the f_1 representation in S1 (Fig. 2a) is transformed in S2 (Fig. 2b) in a dual representation (positive and negative slopes), which is also observed in areas of the frontal lobe (Fig. 2c–f). According to these results, these

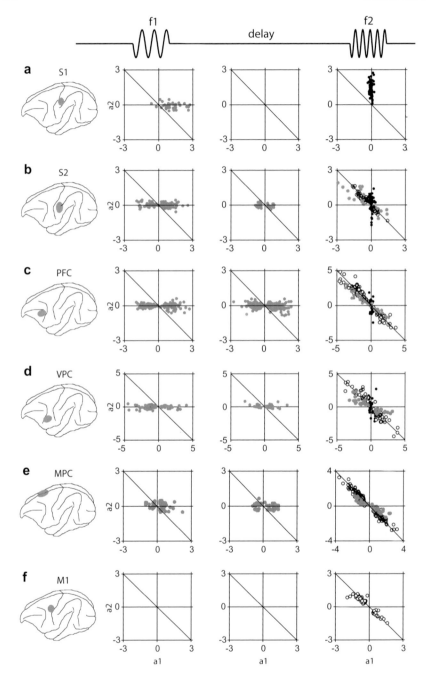

areas of the frontal lobe that process sensory information could also be considered to be parts of the somatosensory system.

Decoding memory processes from neuronal activity

One of the key features of the vibrotactile discrimination task is that it requires short-term memory storage of information about f_1. Because we did not find in S1 any trace of f_1, we wondered where and how the f_1 trace is held in the brain. The first neural correlate about this process was found in PFC (Brody et al 2003, Romo et al 1999), an area involved in working memory. The inferior convexity of the PFC contains neurons that increase their firing rate in a frequency-dependent manner during the delay period (Fig. 2c). The dependence of firing rate on f_1 is monotonic, exactly as it was observed for the f_1 periods in those areas central to S1 (S2, PFC, VPC and MPC). This mnemonic representation is not static, in the sense that the intensity of the persistent activity varies throughout the delay period. Some of the PFC neurons carry information about f_1 during the early component of the delay period, others only during the late part of the delay period, and still others persistently throughout the entire delay period. These findings suggest that in the PFC coexist distinct neuronal populations that carry information of f_1 at different times and may also indicate that the PFC circuit is composed of a chain of neurons that dynamically hold the f_1 information (Brody et al 2003, Miller et al 2003, Romo et al 1999).

FIG. 2.   Population dynamics in different cortical areas during the flutter discrimination task. Each data point represents one neuron. For each neuron, responses were fitted to the equation: firing rate = $a_1 \times f_1 + a_2 \times f_2 + b$. The coefficients a_1 and a_2 were computed from responses at different times during the task. Points that fall on the $a_1 = 0$ axis represent responses that depend on f_2 only (black dots); points that fall on the $a_2 = 0$ axis represent responses that depend on f_1 only (gray dots); points that fall on the $a_2 = -a_1$ line represent responses that are functions of $f_2 - f_1$ (open circles). The data shown are significantly different from (0,0) in at least one of the epochs analysed. (a) S1 responses during the first stimulation period (f_1; left), the interstimulus period (delay; middle), and the second stimulation period (f_2; right). These neurons were active only during stimulation; most of them increased their rates with increasing frequency (positive a_1 and a_2). (b) S2 neurons respond to f_1 (left) and exhibit a modest but significant amount of delay activity (middle). Positive and negative coefficients indicate rates that increase and decrease as functions of frequency, respectively. During the initial part of f_2 (right), neurons may have significant a_1 coefficients (grey dots) or may respond exclusively to f_2 (black dots), as computed from the first 200 ms after stimulus onset. Later on, the coefficients cluster around the line $a_2 = -a_1$ (open circles dots), as computed from the last 300 ms before stimulus offset. Brain diagram shows region of approach to S2, which is hidden in the lateral sulcus. (c–f) Data from prefrontal cortex (PFC), ventral premotor cortex (VPC), medial premotor cortex (MPC) and primary motor cortex (M1) are calculated as in (b). (Modified from Hernández et al 2000, 2002, Salinas et al 2000, Romo et al 2002, 2004, and from unpublished data from Romo et al for PFC and M1.)

An important observation regarding the working memory systems is that other cortical areas also hold information about f_1. The VPC (Fig. 2d) also encodes information about f_1 during the delay period exactly as does the PFC (Romo et al 2004). Also some S2 neurons show a similar type of monotonic encoding (Fig. 2b and 3b), but only during the early part of the delay period, suggesting the presence of working memory signals in S2 (Salinas et al 2000). Whether these S2 neurons are the ones that drive PFC and VPC neurons during the delay period or whether the S2 neurons that respond during the f_1 periods are the ones that activate the mnemonic circuits is not known.

One wonders about this mnemonic coding scheme. Is there any distinction about the functional role of these mnemonic neurons found in these cortical areas? There are a couple of additional observations that may shed light on this question and they came from recordings in the MPC (Hernández et al 2002). First, the MPC contains neurons that encode f_1 during the late part of the delay period, just before the presentation of f_2 (Fig. 3). Again, with similar monotonic responses and similar proportions of positive and negative slopes (Fig. 2e). Second, the dynamics of these neurons are similar to those from PFC and VPC that encode f_1 during the late part of the delay period (Brody et al 2003, Romo et al 1999, 2004). This would suggest a coding mnemonic scheme according to the task demands. Information about f_1 must be available during the f_2 period for the comparison with the f_2 input and persistent and late neurons might provide it (Fig. 3). Persistent and late neurons are therefore well positioned to compute the comparison process.

Decoding comparison processes from neuronal activity

Reaching a decision in the vibrotactile discrimination task requires the comparison between the memory trace of f_1 and the current sensory input of f_2. We sought evidence of this operation in S1, but as indicated already the activity of these neurons do not combine f_1 and f_2 during the comparison period; they encode only information of f_2. Where and how is this neuronal operation executed? A simple inspection of the neuronal activity in areas central to S1 indicated that the responses during the f_2 period are quite complex (Figs. 2 and 3). For example, some S2 neurons encoded f_2 in their firing rates similarly as for f_1 (positive and negative slopes). But, surprisingly, many S2 neurons responded differentially during the comparison $f_2 > f_1$ or $f_2 < f_1$ trials during correct discriminations (Romo et al 2002). These differential responses were even more abundant in areas of the frontal lobe (PFC, VPC, MPC and M1) examined in this task (Hernández et al 2002, Romo et al 2004). The question is whether the responses during f_2 depended on f_1, even though f_1 had been applied 3 seconds earlier, or whether they simply reflected their association with the motor responses. We ruled out the presence of a simple differential motor activity associated with the push-button presses by testing these neurons in a control

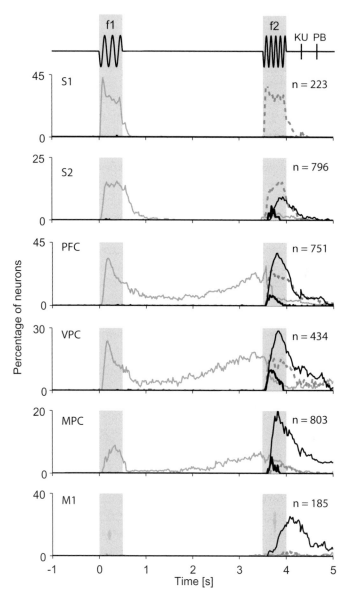

FIG. 3. Dynamics of population responses of six cortical areas during the vibrotactile discrimination task. Coefficient values of a_1 (continous grey trace), a_2 (dotted grey trace), and as a function of the interaction between a_1 and a_2 (continous black traces indicate those neurons that show $f_2 > f_1$ or $f_2 < f_1$; thick black traces indicate those neurons that show the actual difference between f_2 and f_1). The responses are expressed as percentage of the total number of neurons (n) that had task-related responses. S1, primary somatosensory cortex; S2, secondary somatosensory cortex; PFC, prefrontal cortex; VPC, ventral premotor cortex; MPC, medial premotor cortex; M1, primary motor cortex. Original data from S1, S2, MPC, VPC and M1 were previously published (Hernández et al 2000, 2002, Romo et al 2002, 2004) and data from PFC are unpublished results.

task where the same vibrotactile stimuli were used, but animals had to follow a visual cue to produce the motor responses. In this condition all neurons reduced the differential activity (Fig. 4), indicating that the differential activity observed during the comparison period depends on the actual computation between f_1 and f_2 and does not reflect a purely motor response aimed to press one of the two push-buttons (Hernández et al 2002, Romo et al 2002, 2004).

If the neuronal discharges during the comparison period are the product of the interaction between f_1 and f_2, then the trace of f_1 and the current f_2 could be observed during the comparison period before the discharges indicated the motor decision responses. To further quantify these interactions between f_1 and f_2 and beyond it, we used the multivariate regression analysis described already. The analysis revealed the contributions of f_1 and f_2 during the comparison period for S2, PFC, VPC and MPC neurons (Fig. 2). This is clearly shown in the successive time windows by plotting the coefficients a_1 and a_2 and the absolute difference between the two $(a_1 = -a_2)$ during the entire sequence of the vibrotactile task (Fig. 3). This allows appreciating the time dynamics of the neurons' response dependence on f_1 and f_2 for each of the cortical areas that are active during the vibrotactile discrimination

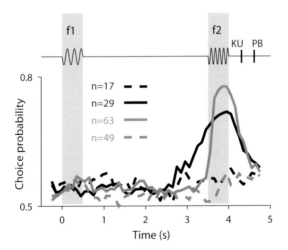

FIG. 4. Correlation between neuronal and behavioural responses. Choice probability indices for ventral premotor cortex (VPC) neurons as functions of time for three different groups of neurons. Results are averaged over (f_1, f_2) pairs. Black broken lines, responses that depended only on $f2$ during the comparison period. Black trace, neuronal responses that depended on $f1$ during the delay period and on $f_2 - f_1$ during the comparison period. Grey trace, neuronal responses that depended on $f_2 - f_1$ during the comparison period, but were not significant during the delay period between f_1 and f_2. Grey broken trace, neuronal responses that had large choice probability indices (black and grey continuous traces), but tested in a control task in which animals had to follow a visual cue to produce the motor response. (Modified from Romo et al 2004.)

task. This comparative analysis shows that the decision-making process is widely distributed through the cortex, although with various strengths across these areas (Romo et al 2004). The comparison signal evolves into a signal, which is consistent with the motor choice, but this is again stronger in some areas than in others, but it is widespread nonetheless. The resulting motor signal is also observed in M1, but M1 does not seem to participate in the sensory, memory and comparison components of the task (Fig. 3). Also, the differential signal in M1 is considerably delayed in comparison to S2, PFC, VPC and MPC (Romo et al 2004). The results suggest that the comparison between stored and ongoing sensory information takes place in a distributed fashion. But, do these neurons predict in their activity the motor decision report?

Decoding decision processes from neuronal activity

Responses during correct trials alone do not allow us to determine to what extent the comparison dependent responses observed in S2 and frontal lobe are correlated with the sensory evaluation, or with the monkey's action choice itself (Figs. 2, 3). To answer these questions, for each neuron we sorted the responses into hits and errors and calculated a choice probability index (Green & Swets 1966, Britten et al 1996, Hernández et al 2002, Romo et al 2002, 2004). This was quantified for each $f_2 - f_1$ pair whether responses during error trials were different from responses during correct trials (Fig. 4). If the responses were exclusively stimulus dependent, they should show little or no difference between error and correct trials. In contrast, if the responses were linked to the monkey's choice, then the responses should vary according to which button the monkey chose to press. In principle, this represents the probability with which an observer of a neuron's response to a given (f_1, f_2) pair would accurately predict the monkey's choice. We found that the closer a neuron's responses to correct trials were purely $f_2 - f_1$ dependent, the higher the separation between responses to correct and error trials, as quantified by a higher choice probability. We also found that the choice probability indices increased during the course of the f_2 period. This was quite evident for those neurons that had $f_2 - f_1$ responses but not for those neurons that responded to f_2 only. This tendency was observed for each area examined central to S1 (Romo et al 2004). We illustrate these processes for subgroups of VPC neurons (Fig. 4). An interesting finding was that the neuronal population that carried f_1 information during the delay period also shows large choice probability (above 0.5) values just before the comparison period (Fig. 4). We suggest that this activity is related to the working memory component of the task as opposed to the decision component of the task. If trial-by-trial variations of f_1 encoding during the working memory period correlate with trial-by-trial variations in performance, this will then be reflected in the choice probability index. The choice probability analysis shows that responses from

S2 and the frontal cortex reflect the active comparisons between f_1 and f_2 and the motor choice that is specific to the context of the vibrotactile discrimination task.

Generating artificial percepts by cortical microstimulation

We have shown how the neuronal activity from several cortical areas is associated with the different components of the vibrotactile discrimination task. But, do these neuronal correlates actually have a direct impact in the task? Intracortical microstimulation is a powerful technique that can be used to directly prove whether the activity of localized groups of neurons is causally linked to the cognitive components of this task (Salzman et al 1990). This approach can be used to test whether the S1 representation of the stimuli is sufficient to trigger all the cognitive processes of the task (Romo et al 1998, 2000). Previous experiments had shown that the quickly adapting (QA) neurons of S1 encode the stimulus frequency both in their periodicity and firing rates (Mountcastle et al 1969). However, an analysis of the neural activity using detection theoretic analysis showed that the firing rate representation correlated with the monkey's psychophysical performance (Hernández et al 2000, Salinas et al 2000). Thus, the microstimulation approach may be useful to test which of these two codes (periodicity or firing rates) are meaningful for discrimination in this task.

Figure 5 summarizes results from several microstimulation experiments. The experiments were carried out using the following protocols: in half of the trials, microstimulation of area 3b of S1 substituted for the mechanical, comparison stimulus frequency (Romo et al 1998). Artificial stimuli consisted of periodic bursts

FIG. 5. Psychophysical performance in frequency discrimination with natural, mechanical stimuli delivered to the fingertips and with artificial, electrical stimuli delivered directly to S1 neurons. Monkeys were first trained to compare two mechanical stimuli presented sequentially on the fingertips (Fig. 1). Then some of the mechanical stimuli were replaced by trains of electric current bursts microinjected into clusters of QA neurons in area 3b. Each burst consisted of two biphasic current pulses. Current bursts were delivered at the same comparison frequencies as natural stimuli. In half of the trials the monkeys compared two mechanical vibrations delivered on the skin; in the other half one or both stimuli were replaced by microstimulation. The two trial types were interleaved, and frequencies always changed from trial to trial. The diagrams on the left show four protocols used. The curves on the right show the animals' performance in the situations illustrated on the left. Filled and open circles indicate mechanical and electrical stimuli, respectively; continuous lines are fits to the data points. (a) All stimuli were periodic; the comparison stimulus could be either mechanical or electrical. (b) The base stimulus was periodic and the comparison aperiodic; the comparison could be mechanical or electrical. (c) All stimuli were periodic; the base stimulus could be mechanical or electrical. (d) All stimuli were periodic; in microstimulation trials both base and comparison stimuli were artificial. Vibrotactile stimuli were either sinusoids or trains of short mechanical pulses, each consisting of a single-cycle sinusoid lasting 20 ms. Monkeys' performance was practically the same with natural and artificial stimuli. (Modified from Romo et al 1998, 2000.)

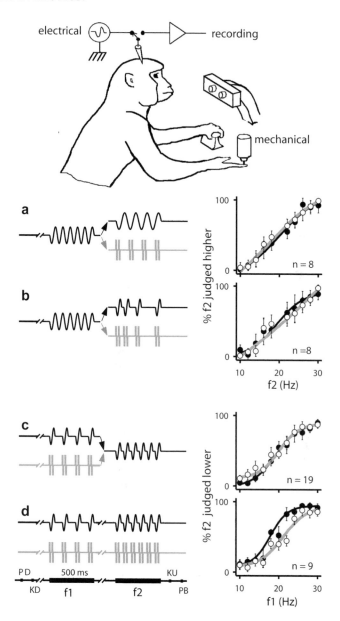

delivered at the comparison frequencies as the mechanical stimuli. Microstimulation sites in area 3b were selected to have QA neurons with small receptive fields on the fingertips at the location of the mechanical stimulating probe. Remarkably, the monkeys could discriminate between the mechanical (f_1) and the electrical f_2 (comparison) stimuli with performance profiles indistinguishable from those obtained with natural periodic stimuli (Fig. 5a). Similar performance levels were observed with aperiodic stimuli (Fig. 5b). The most direct interpretation of these findings is that the electrical stimuli induced sensations that closely resembled those induced by the mechanical stimuli, and that periodicity does not matter for discrimination. This latter observation is consistent with the fact that neurons from areas central to S1 encode the stimuli in their firing rates and that their discrimination thresholds calculated from their firing rate match the psychometric thresholds (Hernández et al 2000, Salinas et al 2000).

Due to the task design, comparison of f_2 is made against the memory trace of f_1. We wondered whether in addition to using artificial stimuli during the comparison period of the task, monkeys could store and use a quantitative trace of an electrical stimulus delivered to clusters of QA neurons in S1 in place of the f_1 mechanical stimulus. We also wondered whether monkeys could perform the entire task on the basis of purely artificial stimuli (Romo et al 2000). This would demonstrate that activation of the QA circuit of S1 was sufficient to initiate the entire cognitive process involved in the task. In experiments in which the f_1 consisted of electrical microstimulation, the monkey's psychophysical performance was again indistinguishable from that quantified with only natural stimuli, indicating that the signals evoked by mechanical and artificial stimuli could be stored and recalled with the same fidelity (Fig. 5c). Moreover, monkeys could perform the entire task, with little degradation in performance, on the basis of purely artificial stimuli (Fig. 5d).

As for substituting the comparison stimulus with electrical patterns, monkeys could not reach the usual level of performance when clusters of slowly adapting (SA) neurons were microstimulated (Romo et al 2000). Nor could they discriminate when microstimulation patterns were made at the border between QA and SA clusters. These control experiments tell us about the specificity of the QA circuit in this task (Romo et al 2000). This indicates that microstimulation elicits quantitative discriminable percepts, and shows that activation of the QA circuit of S1 is sufficient to initiate the entire subsequent neural process associated with vibrotactile discrimination. Relevant to interpreting the S1 microstimulation results, previous studies have shown that activity in a single cutaneous afferent fibre could produce localized somatic sensations (Johansson & Vallbo 1979), and frequency microstimulation of QA afferents linked to Meissner's corpuscles produced the vibrotactile sensation (Ochoa & Torebjork 1983). These observations strongly support the notion that the activity initiated in specific mechanoreceptors is read out by S1; this reading is then widely distributed to those anatomical structures that are linked to S1. The whole sequence of events associated with this sensory discrimination task must depend on this distributed neural signal as already discussed in previous sections.

General discussion

The evidence reviewed here suggests that the comparison between stored and ongoing sensory information takes place in a distributed fashion. It also suggests that there is a continuum between sensory-and motor-related activity. For example, f_1 is encoded in multiple cortical areas. Such encoding seems to proceed in a serial fashion from S1 to S2, PFC, VPC and then to MPC. Although the strength of this signal varies across these areas, all of them except for S1 store f_1 at different times during the working memory component of the task. This is consistent with the proposal that there is a large cortical network that dynamically stores sensory information during working memory (Fuster 1997). During the comparison period, f_2 is processed similarly by the same cortical areas. The comparison between the stored sensory information of f_1 and the current sensory input of f_2 is observed in S2, PFC, VPC and MPC, again with various strengths across the cortical areas. This comparison signal evolves into a signal that is consistent with the motor choice; this is again stronger in some areas than in others, but is widespread nonetheless. The resulting motor signal is also observed in M1, but it does not seem to participate in the sensory, memory, and comparison components of the task.

This comparative analysis shows that in the vibrotactile task, S1 is predominantly sensory and M1 is predominantly motor, but otherwise there is broad overlap in response characteristics across all other cortical areas studied. The difference between S2, PFC, VPC and MPC might best be characterized as shifts in the distributions of response types (Figs. 3, 4). For example, compare PFC, VPC and MPC: their response latencies were significantly different, with the f_1 and f_2 signals beginning slightly earlier in PFC and VPC than MPC (Romo et al 2004). The percentages of neurons that encoded each component of the discrimination task were also different. These findings suggest that the premotor areas may coordinate the sensory, memory and decision components of the task but that these processes are first coordinated in PFC and VPC. This result, however, should be interpreted cautiously, since recordings were made in different animals and the same population from each cortical area may vary from animal to animal.

An interesting finding worth detailed discussion is the existence of neural populations with opposite responses—or, more precisely—of populations with opposite-sign tuning curves (positive and negative slopes). One of the simplest ways in which neurons could encode the frequency of vibratory stimuli is by means of a tuning curve in which particular firing rate values encoded particular stimulus frequencies, determined by any arbitrary function (Romo et al 2003). Then, if all neurons of a given area had similar responses, pooling of individual responses could provide an accurate estimate of the stimulus frequency (the fidelity of this estimate would be determined by the correlation values among neurons; Shadlen & Newsome 1998). Instead of this simple coding scheme, the results showed that, in all areas central to S1, there is not a single, but a dual stimulus encoding. Given that the slopes are of opposite signs (antagonistic responses), pooling the activity of these two groups of neurons would not give any useful information about the stimulus frequency. Therefore, well-structured cortical circuits are necessary to keep the information of each separated population. As we have seen, this dual encoding is preserved along the processing levels, from S2, PFC, VPC and MPC. What is the role of this dual representation?

It has been shown that responses of individual S2 neurons provide less information about the stimulus frequency than individual responses of S1 neurons (Salinas et al 2000, Romo et al 2003). Unlike S1, where the information provided by individual neurons is enough to explain the monkeys' discrimination thresholds, neurometric curves obtained from individual responses of S2 neurons are well below the discrimination thresholds of monkeys (Romo et al 2003). Is sensory information degraded as it flows from S1 to S2? At first sight, this may seem to be the case. However, combining the responses of neurons with opposite slopes could compensate for the loss of information. Indeed, we have shown that it is possible to recover the information apparently lost between S1 and S2 by means of a subtraction operation between pairs of neurons with opposite tuning curves (Romo et al 2003). This operation, which can be thought of as a contrast enhancement mechanism, is particularly useful when neurons show positive correlation coefficients: subtracting the activity of two positively correlated neurons cancels correlated random modulations. Thus, the existence of neuronal populations with opposite signs constitutes a mechanism for representing sensory information along the successive processing stages of cortex, even though significant levels of positive correlation exist among the activity of the neurons. Importantly, this encoding scheme has also been found in the cortices of monkeys that require behavioural decisions based on sensory evaluation (Sinclair & Burton 1991, Freedman et al 2001).

Concluding remarks

The highly simplified sensory discrimination task used here requires perceiving a stimulus, storing it in working memory, combining the stored trace with the current sensory stimulus and producing a decision which is communicated to the motor apparatus. The entire sequence of the task is reflected in the activity of neuronal populations from several cortical areas of the parietal and frontal lobes. Our results indicate that neurons from areas central to S1 do not simply wait for a signal encoding decision, but participate at every step of its generation by combining working memory and sensory inputs. This process is carried out by two complementary neuronal responses. This dual representation is found in all areas central to S1 examined in this task, and might serve to compute optimally the entire perceptual process of the task. This coding scheme has also been found in some cortices of monkeys performing tasks that require behavioural decisions based on a comparison operation. An important problem posed by these findings is whether each neuronal correlate found in each cortical area actually has an impact in the perceptual task. Perhaps, microstimulation experiments of the type carried out in S1 are necessary to prove whether this is so.

Acknowledgements

The research of R. Romo was partially supported by an International Research Scholars Award from the Howard Hughes Medical Institute, Grants from the Millennium Science Initiative-CONACYT and DGAPA-UNAM.

References

Britten KH, Newsome WT, Shadlen MN, Celebrini S, Movshon JA 1996 A relationship between behavioral choice and the visual responses of neurons in macaque MT. Vis Neurosci 13:87–100

Brody CD, Hernández A, Zainos A, Romo R 2003 Timing and neural encoding of somatosensory parametric working memory in macaque prefrontal cortex. Cereb Cortex 13:1196–1207

Burton H, Fabri M, Alloway K 1995 Cortical areas within the lateral sulcus connected to cutaneous representations in areas 3b and 1: a revised interpretation of the second somatosensory area in macaque monkeys. J Comp Neurol 355:539–562

Cipolloni PB, Pandya DN 1999 Cortical connections of the frontoparietal opercular areas in the rhesus monkey. J Comp Neurol 403:431–458

Disbrow E, Litinas E, Recanzone GH, Padberg J, Krubitzer L 2003 Cortical connections of the second somatosensory area and the parietal ventral area in macaque monkeys. J Comp Neurol 462:382–399

Freedman DJ, Riesenhuber M, Poggio T, Miller EK 2001 Categorical representation of visual stimuli in the primate prefrontal cortex. Science 291:312–316

Fuster J 1997 Network memory. Trends Neurosci 20:451–459

Godschalk M, Lemon RN, Kuypers HG, Ronday HK 1984 Cortical afferents and efferents of monkey postarcuate area: an anatomical and electrophysiological study. Exp Brain Res 56:410–424

Green DM, Swets JA 1966 Signal detection theory and psychophysics. Wiley, New York

Hernández A, Salinas E, Garcia R, Romo R 1997 Discrimination in the sense of flutter: new psychophysical measurements in monkeys. J Neurosci 17:6391–6400

Hernández A, Zainos A, Romo R 2000 Neuronal correlates of sensory discrimination in the somatosensory cortex. Proc Natl Acad Sci USA 97:6191–6196

Hernández A, Zainos A, Romo R 2002 Temporal evolution of a decision-making process in medial premotor cortex. Neuron 33:959–972

Hubel DH, Wiesel TN 1998 Early exploration of the visual cortex. Neuron 20:401–412

Johansson RS, Vallbo AB 1979 Detection of tactile stimuli: thresholds of afferent units related to psychophysical thresholds in the human hand. J Physiol 297:405–422

Miller P, Brody CD, Romo R, Wang XJ 2003 A recurrent network model of somatosensory parametric working memory in the prefrontal cortex. Cereb Cortex 13:1208–1218

Mountcastle VB, Talbot WH, Darian-Smith I, Kornhuber HH 1967 Neural basis of the sense of flutter-vibration. Science 155:597–600

Mountcastle VB, Talbot WH, Sakata H, Hyvarinen J 1969 Cortical neuronal mechanisms in flutter-vibration studied in unanesthetized monkeys. Neuronal periodicity and frequency discrimination. J Neurophysiol 32:452–484

Mountcastle VB, Steinmetz MA, Romo R 1990 Frequency discrimination in the sense of flutter: psychophysical measurements correlated with postcentral events in behaving monkeys. J Neurosci 10:3032–3044

Newsome WT, Britten KH, Movshon JA 1989 Neuronal correlates of a perceptual decision. Nature 341:52–54

Ochoa J, Torebjork E 1983 Sensations evoked by intraneural microstimulation of single mechanoreceptor units innervating the human hand. J Physiol 342:633–654

Pons TP, Garraghty PE, Friedman DP, Mishkin M 1987 Physiological evidence for serial processing in somatosensory cortex. Science 237:417–420

Romo R, Salinas E 2001 Touch and go: decision mechanisms in somatosensation. Annu Rev Neurosci 24:107–137

Romo R, Salinas E 2003 Flutter discrimination: neural codes, perception, memory and decision making. Nat Rev Neurosci 4:2032–2018

Romo R, Brody CD, Hernández A, Lemus L 1999 Neuronal correlates of parametric working memory in the prefrontal cortex. Nature 399:470-473

Romo R, Hernández A, Zainos A, Salinas E 1998 Somatosensory discrimination based on cortical microstimulation. Nature 392:387–390

Romo R, Hernández A, Zainos A, Brody CD, Lemus L 2000 Sensing without touching: psychophysical performance based on cortical microstimulation Neuron 26:273–278

Romo R, Hernández A, Zainos A, Lemus L, Brody CD 2002 Neuronal correlates of decision-making in secondary somatosensory cortex. Nat Neurosci 5:1217–1225

Romo R, Hernández A, Zainos A, Salinas E 2003 Correlated neuronal discharges that increase coding efficiency during perceptual discrimination. Neuron 38:649–657

Romo R, Hernández A, Zainos A 2004 Neuronal correlates of a perceptual decision in ventral premotor cortex. Neuron 41:165–173

Salinas E, Hernandez A, Zainos A, Romo R 2000 Periodicity and firing rate as candidate neural codes for the frequency of vibrotactile stimuli. J Neurosci 20:5503–5515

Salzman CD, Britten KH, Newsome WT 1990 Cortical microstimulation influences perceptual judgments of motion direction. Nature 346:174–177

Schall JD 2001 Neural basis of deciding, choosing and acting. Nat Rev Neurosci 2:33–42

Shadlen MN, Newsome WT 1998 The variable discharge of cortical neurons: implications for connectivity, computation, and information coding. J Neurosci 18:3870–3896

Sinclair RJ, Burton H 1991 Neuronal activity in the primary somatosensory cortex in monkeys (Macaca mulatta) during active touch of textured surface gratings: responses to groove width, applied force, and velocity of motion. J Neurophysiol 66:153–169

Talbot WH, Darian-Smith I, Kornhuber HH, Mountcastle VB 1968 The sense of flutter-vibration: comparison of the human capacity with response patterns of mechanoreceptive afferents from the monkey hand. J Neurophysiol 31:301–334

DISCUSSION

Derdikman: According to some theories, when you get to M1 you develop a population vector. Can you think of a population vector already existing in S2, for instance?

Romo: It may be possible, but our push buttons are 2 cm apart. The population vectors described in M1 do not apply to our experimental conditions. Our differential responses are related to the discrimination outcomes, rather than to motor outputs. In fact, the differential responses observed during the discrimination task are no longer present when animals perform push-button presses in a variant of the discrimination task that does not require somatosensory discrimination.

Brecht: This negative tuning is something new to the secondary cortex. It isn't seen in primary cortex. Is it seen in naïve animals?

Romo: Yes, S2 transforms the S1 positive tuning into a dual representation reflected in two separate neuronal populations: one that shows positive tuning and another that shows negative tuning. These two neuronal populations are necessary for the discrimination process (Romo et al 2003). They compute a subtraction operation that enhances sensory information which matches the animal's discrimination threshold. This neural operation, I believe, is forged during the learning process. Of course, this is difficult to prove because we have never tried it in naïve animals. This could be explored with chronic recording techniques, a problem which is currently addressed in our laboratory.

Brecht: In an anaesthetized monkey, do you see negative tuning in S2?

Romo: Most of the tuning responses beyond S1 are manipulated by the behavioural context. When monkeys do push-motor presses during a visual instruction task and the same stimuli are delivered to the fingertips, the tuning responses of S2 neurons are dramatically affected. I would assume that in anaesthesia these tuning responses are lost. However, we have not tested this.

Dehaene: Did you try to manipulate the duration of the stimuli to see whether it is frequency, number or duration that is important?

Romo: Yes and the results suggest that it is the firing rate. By analysing the responses of single neurons recorded in primary somatosensory cortex while trained monkeys discriminated between two consecutive vibrotactile stimuli, we tested five possible candidate codes. Monkeys could discriminate the difference in the frequency of two stimuli by measuring: (1) the time intervals between spikes; (2) average spiking rate during each stimulus; (3) absolute number of spikes elicited by each stimulus; (4) average rate of production of bursts of spikes; or (5) absolute number of spike bursts elicited by each stimulus. We found that each of these codes carries sufficient information about stimulus frequency to account for psychophysical performance. However, only a spike count code where spikes are integrated over a time window that has most of its mass in the first 250 ms of each stimulus period, covaried with behaviour on a trial-by-trial basis; this was consistent with psychophysical biases induced by manipulation of stimulus duration; and produced neurometric discrimination thresholds similar to behavioural psychophysical thresholds. We also know that firing rate is the only code for sensory discrimination in areas central to S1. Therefore, firing rate seems to be the neural code to solving the vibrotactile discrimination.

The sensory component has an early latency that proceeds in a serial fashion, but the comparison component is longer relative to the pure sensory processing. The comparison component, by definition, is the result of the combination of the current sensory input with the working memory component. The latency of this operation is in the order of 180 ms, and that varies as a function of discrimination process.

Harris: You showed the functions for the S2 neurons, relating the frequency of the vibration to the firing rate response. It was basically linear in each case, whether they were ascending or descending. The comparison process seemed to be characterized by a subtraction.

Romo: That is the operation we propose. We have published these results in *Neuron* (Hernández et al 2002, Romo et al 2004) and in *Nature Neuroscience* (Romo et al 2002).

Harris: Would that imply that the discrimination ability should be invariant with respect to base frequency, comparing 12 against 14 would be the same as 22 against 24? In other words, you won't get a Weber fraction, but a flat function.

Romo: It's possible. I have done that comparison as a function of the difference of the frequency of the second and the first, and found two different functions. One that shows the difference between the first and second, then they become very categorical. And another function that simple tells you the categorical response. The first function must be more related to the sensory decision component, whereas the second one is more likely related to the decision motor report.

Harris: If you varied the difficulty, would this have an effect?

Romo: If we compare 20 Hz versus 20 Hz, it is very difficult to see this comparison signal in S2. But, if the difference between the two stimuli is 4 Hz, neurons reflect this. The sensory decision signal grows as a function of the difference between the two stimuli.

Harris: If you had 20 versus 22 and 20 versus 28, would there be much difference.

Romo: As I've already said, some neurons reflect the comparison discrimination process.

Porro: Do positively or negatively tuned cells tend to cluster in different columns?

Romo: We have some ideas in S2. We have already seen those two populations, which seem to be arranged in clusters of neurons and tried to manipulate them through microstimulation as we did for S1. The result we found was quite strange. In microstimulation of a cluster of neurons that showed negative tuning, no matter what stimulus is used the animal has a tendency to report that the second stimulus was lower than the first. In our work, when a monkey makes three or four errors in a row, they change their strategy. Now, during microstimulation in a cluster of neurons with positive slope, we observed exactly the opposite. This is very strange, and I haven't been able to pursue this problem. The recording and microstimulation observations suggest that S2 is organized in clusters of neurons that may correspond to what are called 'columns' in S1. We should focus on this problem.

Porro: There are suggestions from human studies that some kinds of somatosensory discrimination tasks involve the posterior parietal cortex. Have you looked there?

Romo: I simply showed the results obtained for a few cortical areas. We have recorded from a lot more. In brief, information from S1 is widely distributed in the parietal lobe. Most of the neurons from parietal somatosensory fields have the tendency to encode the basal stimulus with the type of responses we observed in S2. Of course, there are many neurons that simply fire to the stimulus and do not encode a stimulus feature. We don't know the contribution of these neurons to the frequency discrimination task.

Porro: Are neurons in the posterior parietal cortex less precisely tuned?

Romo: No. There are neurons from area 7B, 5 or 2 that can be as good as the neurons from S2. However, neurons with no tuning functions can be also observed in S1.

Rizzolatti: At least from this plot, the behaviour of ventral premotor neurons seems to be very similar to S2 neurons.

Romo: The plot shows the relevant response during the base, working memory and comparison periods. The difference between S2 and ventral premotor cortex is the working memory component. S2 neurons are tuned to base stimulus, then the response is prolonged to the delay period between the two stimuli for about 500 ms. In the ventral premotor cortex you can have tuning during the base stimulus that is preserved along the whole delay period. There are some neurons that do not respond to the base stimulus, but they start to fire just before the second stimulus. It may be an expectation signal according to psychologists, but when we do the analysis we observe a gradient response linearly related to the base stimulus. Therefore, there is substantial difference in the response patterns of neurons from S2 and premotor cortices and of course with primary motor cortex.

Schall: The latency of the comparison response struck me. Are any of the cells responding early in an undifferential manner? Or are they differential from the beginning of the response?

Romo: The beginning of the differential response is about 200 ms. This is more or less the type of response you have measured in the frontal eye field. Apparently, there is something common which might be independent of the sensory modality. If I were there in the VPC with a visual task I probably could see this type of response. I want to believe that these neurons might respond to many sensory modalities, but it is very difficult to test the animals with more than one task.

Schall: If it is delayed according to the difficulty of the discrimination, then the latency of the response should be delayed when the stimuli are closer together.

Romo: We can measure this in the comparison process, but during the comparison process we have sensory responses with very short latencies. The differential responses grow as a function of time during the comparison period.

Logothetis: Were these recordings from different areas done simultaneously?

Romo: We are now trying this by recording simultaneously neurons in S2 and frontal cortex.

Logothetis: If you have simultaneous recordings it will be interesting to apply some kind of modelling, to see whether from two or three neurons tuned to two different frequencies you have a good way to predict what is happening with cells at the single spike level.

Romo: We are analysing the data collected with the two implanted apparatuses.

References

Hernández A, Zainos A, Romo R 2002 Temporal evolution of a decision-making process in medial premotor cortex. Neuron 33:959–972

Romo R, Hernández A, Zainos A, Lemus L, Brody CD 2002 Neuronal correlates of decision-making in secondary somatosensory cortex. Nat Neurosci 5:1217–1225

Romo R, Hernández A, Zainos A, Salinas E 2003 Correlated neuronal discharges that increase coding efficiency during perceptual discrimination. Neuron 38:649–657

Romo R, Hernández A, Zainos A 2004 Neuronal correlates of a perceptual decision in ventral premotor cortex. Neuron 41:165–173

Probabilistic mechanisms in sensorimotor control

Konrad P. Körding and Daniel M. Wolpert[1]

Sobell Department of Motor Neuroscience, Institute of Neurology, University College London, Queen Square, London WC1N 3BG, UK

Abstract. Uncertainty constitutes a fundamental constraint on human sensorimotor control. Our sensors are noisy and do not provide perfect information about all the properties of the world. Moreover, our muscles generate noisy outputs and many tasks we perform vary in an unpredictable way. Here we review the computations that the CNS uses in the face of such sensory, motor and task uncertainty. We show that the CNS reduces the uncertainty in estimates about the state of the world by using a Bayesian combination of prior knowledge and sensory feedback. It is shown that these mechanisms generalize to state estimation of ones own body during movement. We review how the CNS optimizes decisions based on these estimates, examining the error criterion that people optimize when performing targeted movements. Finally, we describe how signal-dependent noise on the motor output places constraints on performance. Goal-directed movement arises from a model in which the statistics of our actions are optimized. Together these studies provide a probabilistic framework for sensorimotor control.

2005 Percept, decision, action: bridging the gaps. Wiley, Chichester (Novartis Foundation Symposium 270) p 191–202

The framework of Bayesian decision theory has emerged as a principled approach to determine optimal behaviour in the face of uncertainty. It combines knowledge about rewards and losses with Bayesian statistics. Numerous articles summarize the mathematical (Cox 1946) and philosophical (Freedman 1995) ideas that are behind Bayesian statistics and how to use them to construct optimal systems (MacKay 2003) for example for navigating robots (Thrun 2000). Bayesian statistics, however, is not only employed to build optimally behaving technical systems but is also moving towards being a major driving factor in understanding the brain. Recently, evidence has been mounting that the behaviour of the human sensorimotor and cognitive systems are captured remarkably well by Bayesian models, with evidence coming from the visual system (Weiss et al 2002, Freeman 1994, Kersten & Yuille

---

[1] This paper was presented at the symposium by Daniel Wolpert, to whom correspondence should be addressed.

2003), the integration of proprioception and vision (Van Beers et al 2002, Ernst & Banks 2002), our cognitive ability to learn concepts from a few examples (Tenenbaum & Griffiths 2001), to model the causal relations between events (Tenenbaum & Griffiths 2002) and to make judgement of randomness (Griffiths & Tenenbaum 2005). In addition, evidence is accumulating that neural properties reflect such statistical processing of inputs, such as in the primary visual cortex (Sharma et al 2003) and in neurons involved in making decisions (Gold & Shadlen 2002). Moreover, Bayesian theories have also recently been used as the foundation of models of language processing (Manning & Schütze 1999). Bayesian statistics has, therefore, allowed novel questions to be addressed in psychophysical and neurophysiological experiments while providing a coherent framework in which information processing can be understood. The main focus of this chapter is on the concepts behind Bayesian methods that are used to understand the human brain, and particularly those that can unify the processes involved in sensorimotor integration and control.

Bayesian integration for probabilistic tasks

Many tasks have natural variability. For example, consider playing tennis and estimating the location where the ball will bounce (see Fig. 1A). Because vision does not provide perfect information about the ball's location or speed there will be uncertainty about the bounce location. However, if we have had previous experience of tennis we can have prior information about where our opponent is

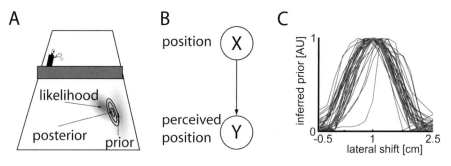

FIG. 1. (A) When an opponent hits a ball into our side of the court we may wish to estimate its bounce location. From vision we can estimate the likelihood of different bounce locations (likelihood region with saturation proportional to likelihood). Prior knowledge may suggest that balls tend to land in another region (prior with saturation proportional to probability). Integrating these two sources of information gives the ellipses that denote the posterior, the region where the Bayesian estimate would predict the ball to land. (B) The true ball's bounce position, x, influences the perceived position, y. (C) In a task in which a parameter (lateral shift) had a prior Gaussian distribution (thick grey line) subject's behaviour allowed their estimate of the prior to be inferred (thin black lines) (Körding & Wolpert 2004).

likely to hit the shot. This prior knowledge can be combined with our estimate based on the visual input to obtain an optimal estimate of where the ball will hit the ground.

This example can be abstracted the following way (Fig. 1B): The physical properties of the ball define the location x where the ball will hit the ground. The visual system, however, does not perceive where the ball will really hit the ground but rather some noisy version of this, y. Given that we perceive the ball at position y, and knowing the uncertainties in the visual system, we can calculate how likely this perception is for different actual bounce locations. This is called the *likelihood* and is denoted $p(y|x)$, that is the probability of perceiving the sensory feedback we have, y, given a hypothesized true location x. We can plot the likelihood for different values of location, x, and this is sketched in Fig. 1A (with saturation proportional to likelihood). Based on the likelihood alone our best estimate would be in the middle of the left hand cloud, that is the maximum likelihood estimate. However, this procedure ignores prior knowledge about the way our opponent plays. In particular, over the course of many matches the positions where the ball hits the ground will not be uniformly distributed, but highly concentrated within the confines of the court and, if our opponent is a good player, this density will be highly peaked near the boundary lines where it is most difficult to return the ball. This probability distribution of positions where the ball hits the ground, $p(x)$, is called the *prior* distribution (shown in Fig. 1a right-hand cloud) and could be learnt using methods of *Bayesian learning*. We can apply *Bayes' Rule* to compute the *posterior*, $p(x|y)$, the probability of the ball being at a particular location x, given we see it at location y.

$$p(x|y) = \frac{\overbrace{p(y|x)}^{Likelihood}\overbrace{p(x)}^{prior}}{p(y)}$$

Note that this formula takes into account both the likelihood and the prior and gives us an estimate of the posterior for each possible bounce location (black ellipses Fig. 1a). This allows us to estimate the most probable bounce location given the prior and sensory information.

If we assume that the prior distribution $p(x)$ is a symmetric two dimensional Gaussian with variance σ_p^2 and mean μ and that the sensory feedback is corrupted with noise so that the likelihood $p(y|x)$ is also a symmetric two dimensional Gaussian with variance σ_V^2 and mean y it is possible to compute the optimal estimate of bounce location $\hat{x} = \alpha y + (1 - \alpha)\mu$ where $\alpha = \sigma_p^2/(\sigma_p^2 + \sigma_V^2)$. It is thus possible to define the optimal estimate given the visual input and prior knowledge. We can also calculate how much better the estimate is compared to a strategy ignoring the prior knowledge. If only the visual feedback is used then variance of the estimate is σ_V^2. However, the Bayesian estimate that incorporates the prior has variance

$\dfrac{\sigma_p^2}{\sigma_p^2 + \sigma_V^2}\sigma_V^2$ which is always less than the variance of the non-Bayesian estimate. If

the prior has the same variance as the sensory feedback then the variance of the Bayesian estimate is half the variance of the non-Bayesian estimate.

In a recent experiment we tested whether people use such a Bayesian strategy (Körding & Wolpert 2004). Subjects had to estimate the value of a one-dimensional variable, the displacement of a cursor relative to the hand, in close analogy with estimating the position where the ball will land. This variable was drawn randomly for each trial out of a Gaussian prior distribution and subjects received extensive training so that they could learn the prior. In addition, on each trial they received brief visual feedback from which they could estimate the likelihood. The experiment allowed us to measure the subject's estimate of the displacement. From these data it is possible to infer the prior that people used. If they ignore the prior information the prior should be flat. The data shown in Fig. 1C (thin black lines) show that people used a prior that was very close to the true prior (thick grey line). This experiment shows that people can use Bayes' rule to estimate the state optimally in a task which varies probabilistically.

State estimation and Kalman controllers

If we move our hand in the dark we have uncertainty about its exact position or velocity because our proprioceptive sensors are not perfect (Fig. 2A). We are thus faced with the problem of estimating the state x of the hand, which is

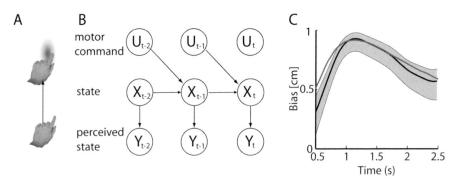

FIG. 2. (A) When we move our hand in the dark, there is uncertainty about its exact location. (B) The state, x, at time t is influenced only by the previous state and the motor command, u. The perceived state, y, is the state with added noise. (C) The error of the subject's estimates of their hand location as a function of the duration of the movement (black line with error lines). The optimal (assuming overestimated force) Kalman controller predicts the grey curve (Wolpert et al 1995).

characterized by its position and velocity. From experience people know how the state of the hand changes over time. In particular they know that the position changes proportional to its velocity and that the velocity change is related to the force applied, although there might be noise in this process. The subjects thus have *prior* knowledge that they can combine with the *likelihood* provided by their sensors. In this case however they constantly have to update their estimate thus effectively using *Bayes' rule* recursively at each time-step.

The model for this process is shown in Figure 2B. Given the real state x of our hand then the perceived state y will be some noisy version of this and u is the motor command we send to our muscles. We know something about the structure of the problem: The state of the hand at time t only depends on the state of the hand at the previous time step, $t-1$ and the applied force, u, this is the *Markov property*. That is the current state of the hand does not explicitly depend on the state of the hand at any but the preceding time. Using *Bayes' rule* twice it is possible to obtain the posterior $p(x_t|x_{t-1}, y_t, u_{t-1}) \approx p(x_t|x_{t-1}, u_{t-1})p(y_t|x_t)$. Here, the estimate is made up of two comments. The first, $p(x_t|x_{t-1}, u_{t-1})$, is the probability of finding oneself in state x_t after having applied command u_{t-1} in state x_{t-1} at $t-1$. To estimate this a *forward model* is used which, by reflecting the dynamics of the arm and hand, can determine the effect of the motor command on this previous state and thereby predict the current state. The second component, $p(y_t|x_t)$, determines how likely the sensory feedback is given this state estimate.

Assuming white noise on the sensory feedback and on the motor commands, the equations obtained describe the optimal *Kalman controller* and it is possible to derive the optimal strategy for predicting the state of the hand: $\hat{x}_t = \hat{A}\hat{x}_{t-1} + \hat{B}u_{t-1} + K_t(y_{t-1} - \hat{x}_{t-1})$ where $\hat{x}$ is the current optimal estimate, $\hat{A}$ is a matrix that characterizes how the hand moves without perturbation, $\hat{B}$ a matrix that characterizes how forces change the state of the hand and K_t is the *Kalman gain* which is a function of the other matrices. The Kalman controller is a generalization of the *Kalman filter* (Kalman 1960) that does not allow a motor signal.

In an experiment, subjects moved their hands in the dark and after each movement they had to estimate where their hand was (Wolpert et al 1995). Movements of varying temporal duration between 500 ms and 2500 ms were performed. Subjects systematically estimated that their hand had moved further than it actually had moved (Fig. 2C, grey). An optimal Kalman controller (Fig. 2C, dark grey) produced very similar results if it was assumed that people systematically overestimate their forces. For small durations the overestimation of distance increased with time. This is because the overestimated force translates into an overestimated distance. However, as the duration increased, the likelihood, which does not depend on the subject's estimate of force, becomes more important compared to the prior. That is why the controller becomes better if the movement lasts a longer period of time. The optimal controller thus shows similar errors to those made by human subjects.

It thus seems that people are able to continuously update their estimates based on information coming in from the sensors using a Bayesian algorithm.

Internal rewards and uncertainty

Only few tasks, such as throwing a dart at a dartboard, explicitly assign a numerical value to the outcome of our movements. However, in the majority of tasks such as pointing, no explicit metric of error is provided although there is the notion that we want to point as precisely as possible. Computationally, any errors we make may be rated by assigning a loss to each error (Fig. 3A). In other words we may internally generate a dartboard that determines how bad an error of 2 cm is compared to an error of 1 cm. An important question is how our errors map onto the loss that is how this internal dartboard looks like. For example, it may be that in pointing only hitting the target counts as good and all other errors are equally bad (Fig. 3B top board). Alternatively the brain may generate internal loss that is proportional to the absolute error, that is distance from the target (Fig. 3B middle board), or quadratically with distance (Fig. 3B bottom board). Measuring the properties of this internal dartboard—that is, how error translates into loss—asks for a combination between statistical theory and information about rewards, called *Bayesian Decision Theory*. It has recently been shown that people use these methods when they move in the presence of uncertainty and monetary rewards (Trommershauser et al 2003). Using experiments where people could control their average pointing errors but not the distribution of errors (Fig. 3C), we could infer the loss function they use for targeting movements (Kording & Wolpert 2004). This showed that the loss

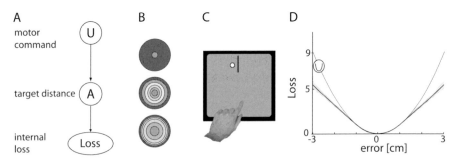

FIG. 3.   (A) Depending on the motor commands there is a probability distribution or distance from the target and each is assigned an internal loss. (B) An important question is how bad an error is compared to another error. Several such virtual dartboards are sketched. (C) White spheres appear on a computer screen the mean position of which is controlled by a hand. The distribution of errors, however, is controlled by a computer program. (D) The loss function inferred in this experiment shown together with a quadratic function (Kording & Wolpert 2004).

function was quadratic for small errors but that it was less than quadratic for large errors making it robust to outliers (Fig. 3D).

The motor system is highly redundant, in that any task can in theory be achieved with infinitely many hand path joint configurations, levels of muscle contraction, etc. Several studies have sought to understand why stereotypy is such a prominent feature of human movement. These studies place motor learning within an optimal control framework in which a task is associated with a cost, and planning or learning can be considered as producing the movement which minimizes the cost. For example, for arm movements several costs which penalise lack of smoothness, either by penalizing the rate of change of acceleration of the hand (jerk) (Flash & Hogan 1985) or the rate of change of torques at the joints (Uno et al 1989) have been successful at modelling human movements. Recently, costs which take into account the variability of the motor command have been developed. Force production shows signal-dependent noise: that is, variability in force with a constant coefficient of variation. Harris & Wolpert (1998) suggested that controlling the statistics in the presence of a signal dependent on the motor command is a major determinant of motor planning. Recently, Todorov & Jordan (2002) have shown that optimal feedback control in the presence of signal-dependent noise may form a general strategy for movement production. This model suggests that rather than form a desired trajectory the motor system may use feedback to control deviations that interfere with the task goal. These models both place noise as a fundamental determinant of human behaviour and have been able to explain a great deal of human planning and control data.

Here we have reviewed the processes of estimation, evaluation through loss functions and optimal control. Taken together these studies suggest a fundamental role of probabilistic mechanisms in sensorimotor control.

Acknowledgements

This work was supported by the Wellcome Trust, McDonnell Foundation and Human Frontiers Science Programme. We thank James Ingram for technical assistance.

References

Cox RT 1946 Probability, frequency and reasonable expectation. Am J Physics 14:1–13
Ernst MO, Banks MS 2002 Humans integrate visual and haptic information in a statistically optimal fashion. Nature 415:429–433
Flash T, Hogan N 1985 The coordination of arm movements: an experimentally confirmed mathematical model. J Neurosci 5:1688–1703
Freedman DA 1995 Some issues in the foundation of statistics. Foundations of science 1:19–83
Freeman WT 1994 The generic viewpoint assumption in a framework for visual perception. Nature 368:542–545
Gold JI, Shadlen MN 2002 Banburismus and the brain: decoding the relationship between sensory stimuli, decisions, and reward. Neuron 36:299–308

Griffiths TL, Tenenbaum JB 2005 From algorithmic to subjective randomness. Presented at Advances in Neural Information Processing Systems, available at *http://books.nips.cc/papers/files/nips16/NIPS2003_CS07.pdf*

Harris CM, Wolpert DM 1998 Signal-dependent noise determines motor planning. Nature 394:780–784

Kalman RE 1960 A new approach to linear filtering and prediction problems. J Basic Eng (ASME) 82D:35–45

Kersten D, Yuille A 2003 Bayesian models of object perception. Curr Opin Neurobiol 13:150–158

Körding KP, Wolpert DM 2004 The loss function of sensorimotor learning. Proc Natl Acad Sci USA 101:9839–9842

Körding KP, Wolpert DM 2004 Bayesian integration in sensorimotor learning. Nature 427:244–247

MacKay DJC 2003 Information theory, inference, and learning algorithms. In: MacKay DJC (ed) Information theory, inference, and learning algorithms. Cambridge University Press, UK also available online at *http://www.inference.phy.cam.ac.uk/mackay/itila/book.html*

Manning CD, Schütze H 1999 Foundations of statistical natural language processing. In: Manning CD, Schütze H (ed) Foundations of statistical natural language processing. The MIT Press, Cambridge, MA

Sharma J, Dragoi V, Tenenbaum JB, Miller EK, Sur M 2003 V1 neurons signal acquisition of an internal representation of stimulus location. Science 300:1758–1763

Tenenbaum JB, Griffiths TL 2001 Generalization, similarity, and Bayesian inference. Behav Brain Sci 24:629–640, 652–791

Tenenbaum JB, Griffiths TL 2002 Theory-based causal inference. Advances in Neural Information Processing Systems, available at *http://books.nips.cc/papers/files/nips15/CS06.pdf*

Thrun S 2000 Probabilistic Algorithms in Robotics. AI Magazine 21:93–109

Todorov E, Jordan MI 2002 Optimal feedback control as a theory of motor coordination. Nat Neurosci 5:1226–1235

Trommershauser J, Maloney LT, Landy MS 2003 Statistical decision theory and the selection of rapid, goal-directed movements. J Opt Soc Am A Opt Image Sci Vis 20:1419–1433

Uno Y, Kawato M, Suzuki R 1989 Formation and control of optimal trajectories in human multijoint arm movements: Minimum torque-change model Biological Cybernetics 61:89–101

Van Beers RJ, Baraduc P, Wolpert DM 2002 Role of uncertainty in sensorimotor control. Philos Trans R Soc Lond B Biol Sci 357:1137–1145

Weiss Y, Simoncelli EP, Adelson EH 2002 Motion illusions as optimal percepts. Nat Neurosci 5:598–604

Wolpert DM, Ghahramani Z, Jordan MI 1995 An internal model for sensorimotor integration. Science 269:1880–1882

DISCUSSION

Scott: People tend to use the robust estimator, but are the results partly reflecting the instruction given to the subject. Have you tried to play with a very specific instruction?

Wolpert: No. If we said you would only get a point for getting it on the target, I don't know whether people could now choose the optimum behaviour which is the mode of the distribution. Certainly, people have shown that if you get explicit rewards in terms of, for example, points, then people are very good at placing their finger at the right place to maximise these rewards.

Gold: Have you looked at different kinds of distributions? What you covered with the skewed distribution seemed like just a small piece of the possible space.

Wolpert: We chose these distributions as with Gaussian distributions all the loss functions predict the same things. There are an infinite number of distributions we could have looked at but we chose these bimodal ones as they were easy to manipulate.

Derdikman: What are the differences between humans and monkeys in performing these tasks?

Wolpert: There are some interesting things in terms of the prediction under signal-dependent noise that arise as the monkey's eye dynamics are very different from those of human eyes.

Logothetis: You apply this under certain conditions and get wonderful answers. If we set all the conditions we get the answer we expect. How general is this? It is easier to imagine the implementation of any Bayesian principle in the brain; it is harder to generalize it for behaviour. The brain in its own environment does have the opportunity to build a huge number of prior distributions. With your behaviour, with the diversity of things you have to do, in many cases you have no clue about the priors. Not only this, but also in many cases it is clear we are constrained by things that have nothing to do with objective priors. You mentioned in your presentation that you minimize movement. This is not true as far as body movement goes. In fact, physiotherapists wouldn't have any work if we did that. It has been shown that in most cases we have paradoxical movements that are based on weak or strong points in the body. If we try to raise our hand and take a book from a library shelf, about 80% of the people do the wrong movement. It is not optimized with respect to anything.

Wolpert: That's a slightly different issue. In terms of the statistics in the real world we have an interest in measuring the statistics of what you do. For example, we know the statistics of visual scenes and auditory stimuli affect the way the brain develops. But no one has ever measured the statistics of what we do with our bodies. We have started implementing an experiment that involves wearing a backpack that measures statistics of what we do every day. We are trying to relate this to what is easy and hard to learn. It turns out that it explains some of the bizarre things that seem to occur. For example, people find it hard to rub their tummy and pat their heads. Why is this? It turns out that if you have a system which has limited resources that it can apply to different tasks and you look at statistics of tasks we have to perform, you begin to understand the way the brain should code motor control. The statistics have a profound effect on the way the brain allocates resources. I am not saying that our movements are optimal: we are not all championship tennis players or Olympic sprinters, so there is variability. One question we can ask is why are some people more skilled than others? There are three options in this model. One is that some people have much more noise and what they end up doing is much more variable. Another is that we don't all find the optimum. Perhaps, if I keep practising I will find the optimum. The third is that we have very different body dynamics. This affects the way that noise propagates through the system. Darts players tend to be very large men. Does having a big body stop the noise going through the system, damping it down? I think we are more optimal than you might think.

Derdikman: I am not sure I understand why a computer plays chess so much better than we do, yet on the other hand tasks which are simple to us are very difficult for computers. The answer you were giving is that it may be because of noise. But if you program a computer to be noisy, it doesn't help it solve tasks we can do.

Wolpert: If I had an infinitely powerful computer, I could write down the trivial algorithm in five lines to solve the chess problem. I would get it to go through possible moves and choose the optimum. The problem is that you need fast computers to achieve this, but the problem itself can be quite easily posed. In motor control you can't pose the question that way and the available space becomes much bigger than with the chess problem. In chess you know exactly where the pieces are, and there are strict limits on the moves they can make. With the body, there are variations in what you get back and variations in what you send out, which makes the problem much harder. This problem is encountered in robotics, and they have noisy sensors and noisy controllers. But they have much less noise than we are exposed to ourselves. It is amazing what we can do given all the noise.

Dehaene: I was wondering whether the strategy for optimum choice applies only to the motor system. Or does it also apply to higher-level processes, such as executive control? In the domain of cognitive arithmetic, Siegler has proposed a strategy choice model that looks very much like what you are suggesting. The idea is that children have a repertoire of strategies for solving arithmetic problems, for instance they can retrieve from memory or try to count on fingers. Children seem to select the choices quite optimally on the basis of their estimates of how fast a given strategy is or how likely it is to achieve success?

Wolpert: In theory, yes. It seems that at the cognitive level people are really badly off and that the motor system is much smarter than the cognitive system. The motor system can solve differential equations, for example. There is some evidence that in cognitive learning, you can explain quite a lot of things on the basis of optimal Bayesian processing.

Understanding Bayes' rule can also explain why we have bizarre perceptions. For example, you might think that visual illusions are the result of sloppy processing, but it turns out that they can be explained by saying that they are the result of optimal processing for the statistics of the world. For example, it can explain some motion illusions by saying that we have a prior probability that things don't move. If we have noise in the visual feedback then some motion illusions fall out as optimal percepts in the face of uncertainty.

Treves: I enjoyed your talk but I am not sure about the explanatory power of invoking Bayesian principles. If what we really needed to understand, in order to produce a computer that can control the movement of chess pieces, is this Bayesian principle, then we would have done it already.

Wolpert: People are starting to do this. Sebastian Thrun is using Bayesian principles now. It is just one part of the solution.

Treves: In a sense, your initial hope of reducing the numbers of papers is not being realized.

Wolpert: I think we are a long way off. That is the hope for the future. Physics textbooks get bigger and then smaller: they oscillate. The hope is that we will end up compacting neuroscience textbooks.

Treves: That's certainly a hope we all share. My fear is that a computationally hard problem in motor control is not in understanding the fact that we have prior expectations, but in describing what those priors look like.

Wolpert: The important thing is learning the dynamics of objects, which is part of the prior, if you take your prior to be knowledge about the world. The suggestion is that we have to represent this separately.

Sparks: What is the physiological evidence consistent with signal-dependent noise in the musculoskeletal system?

Wolpert: People like Schmitt have done a lot of psychological work, where they look at force of impulse while generating constant forces. You can measure the variability of that, and also the fact that the variability goes up with the amplitude of the movement is consistent with signal-dependent noise. We know the properties of the lower motor neurons. They fire with Gaussian inter-spike intervals. If we take this, it isn't enough: it gives you variance going with mean rate. If you put that together with recruitment properties of muscles then you get signal-dependent noise. We have now started looking at how this works in different muscles: it turns out that smaller muscles have much more noise than bigger muscles. If you want to accurately generate 1 Newton, don't do it with your finger, but with your elbow.

Haggard: Memory seems to be involved here. You said it might make sense to make a movement that actually misses the target, because its error ellipse is effectively very narrow. In order for this rather paradoxical result to hold we need some representation of the variability of our own movements. We might have remembered the pattern of errors over the last 1000 movements. I worry about reconciling this with the fact that people seem to know very little about their own movement at a conscious and explicit level.

Wolpert: We may not need a memory. There are other ways we can get this result. Slow learning, which adapts a bit with each trial, effectively implements a memory trace over previous trials. One issue we have been looking at is how many trials are needed for people to learn. It looks as if the statistics of the task can be changed half way through and people slowly change. They seem very good at going from a task with large variance to one with small variance, but have more problems going the other way.

Logothetis: I have a comment on Alessandro Treves' comment about the utility of Bayesian thinking. This is a nice way of seeing certain things and explaining certain strategies. However, I think it would be a big mistake to think it can give computational rules that can be applied in all instances.

Wolpert: I am impressed with how many things it can explain, such as visual illusions. Rather than document 50 illusions as separate entities, it may be possible to explain them all with the prior and the noise. That's an elegant framework.

Logothetis: In real life we don't always have the chance to have an average behaviour. If you are constrained and have to have an answer, this answer may be the ultimate answer you give to the problem.

Wolpert: Absolutely. Bayesian decision making can deal with all these things. Perhaps you don't use it for all these things, but it can certainly deal with ones where we don't know what the prior is and don't know what the noise is.

Human brain activation during viewing of dynamic natural scenes

Uri Hasson and Rafael Malach*

*Center for Neural Science, New York University, New York, NY 10003, USA and *Department of Neurobiology, Weizmann Institute of Science, Rehovot 76100, Israel*

Abstract. To what extent do brains of different human individuals operate in a similar manner? Here we explored the organization and function of different brain regions under progressively more natural conditions. Applying an unbiased analysis, in which spatio-temporal activity patterns in one brain were used to 'model' activity in another brain, we found a striking level of voxel by voxel synchronization between individuals during free viewing of an audio-visual movie. This intersubject correlation was evident not only in primary and secondary visual and auditory areas, but also in association cortices. The results reveal a surprising tendency of individual brains to 'tick collectively' during natural vision. Moreover, our results demonstrate that the unitary nature of conscious experience in fact consists of temporally interleaved and highly selective activations in an ensemble of specialized regions, each of which 'picks-up' and analyses its own unique subset of stimuli according to its functional specialization. Applying reverse correlation to the movie stimuli provides a powerful methodology for revealing both known and unexpected functional specializations in those cortical areas activated by the movie.

2005 Percept, decision, action: bridging the gaps. Wiley, Chichester (Novartis Foundation Symposium 270) p 203–216

A necessary step in establishing a neurobehavioural correlation is to establish the consistency and reliability of the activation pattern across different brains and experimental protocols. For example, a complex, albeit consistent, network of seven functionally distinct object-related regions has been identified along the entire extent of the human occipitotemporal cortex (Grill-Spector 2003, Hasson et al 2003). While the functional properties of these areas are still not fully characterized, it is well established, that these category-related activations reflect a robust phenomenon since the *neuroanatomical* relationships between the functionally specialized regions are highly consistent across subjects and studies (Haxby et al 2001, Hasson et al 2003, Ishai et al 1999). These findings initiated a series of experiments aimed to reveal the underlying organization principles of such consistent activation pattern (for review see Grill-Spector 2003).

   A common theme in these studies is the use of highly controlled visual stimuli. Thus, pre-determined, static and isolated object images are used, which are briefly flashed on the screen during carefully maintained fixation. The use of precisely parameterized stimuli is critical for isolating the experimentally relevant dimensions out of the extremely multidimensional natural stimuli. Furthermore, temporal segregation of stimuli appeared necessary to allow sufficient separation of the sluggish haemodynamic responses from each other.

However, natural vision differs drastically from such highly controlled stimuli along at least four fundamental dimensions: (a) objects are not presented in spatial isolation and are embedded in a complex scene which can include a diversity of object categories; (b) objects move continuously within the scene; (c) subjects freely move their eyes; and (d) seeing usually interacts with additional modalities, as well as with the context and emotional valence of the stimuli. Thus, the world seen in the controlled experimental settings bears little resemblance to our natural viewing experience. Recently we examined to what extent the functional architecture of the human cortex, and in particularly the functional organization of higher-order visual areas, is maintained under more natural conditions (Hasson et al 2004).

We approached this question by studying the functional organization of human cortex under free viewing of a long (30 minutes) sequence taken from an original audio-visual feature film. We reasoned that such rich and complex stimuli are much closer to ecological vision compared to the controlled stimuli usually used in the lab. Recently, a similar approach has been adopted also by Bartels & Zeki (2004a, b).

Obviously, such an open-ended data set is unsuitable for conventional analysis methods (such as the general linear model), due to its spatial and temporal continuity and complexity, multidimensionality, and lack of any prefixed protocol (but see Bartels & Zeki 2004a). We therefore used a new analysis approach:

- In the 'intersubject correlation' analysis we used the timecourse of activation in each voxel of one brain to 'predict' the activity in other fellow brains.
- In the 'reverse correlation' analysis we identified the particular attributes or dimensions of complex natural stimuli that corresponded to peaks in the responses of specific brain areas.

Put simply, our *natural viewing* paradigm inverts the classical approach which uses a set of predefined stimuli to locate brain regions. Instead, we used the brain activations themselves to find the preferred stimuli embedded in the complex stimulation sequence.

The intersubject correlation

Due to the inherent difficulty in employing conventional hypothesis-driven analysis methods in the natural viewing setting we developed a new intersubject correlation measure for searching for brain areas that showed consistent activation patterns across subjects during natural viewing. More specifically, in the inter-subject correlation analysis, we used the spatiotemporal activation profile in a given source brain to predict the activity pattern in other fellows' brains. For that end we normalized all brains into a Talairach coordinate system, spatially smoothed the data, and then used the time-course of each voxel in a given source brain as a predictor of the activation in the corresponding voxel of the target brain (see Fig. 1).

The strength of this across-subject correlation measure stems from the fact that it allows the detection of all sensory-driven cortical areas without the need for any prior assumptions as to their exact functional role. Surprisingly, despite the free viewing and complex nature of the movie we found an extensive and highly significant correlation across individuals watching the same movie. Thus, on average over 29% ± 10 SD of the cortical surface showed highly significant intersubject cor-

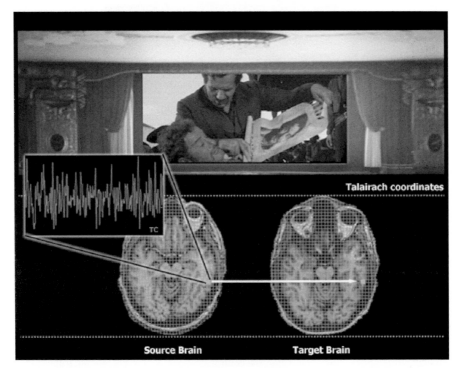

FIG. 1. Voxel-by-voxel intersubject correlation. In this analysis we searched for inter-subject correlation by using the spatiotemporal activation profile in a source brain to predict the activation in other fellow brains. To that end we normalized all brains into a Talairach coordinate system, spatially smoothed the data, and then used the time-course of each voxel in a given source brain as a predictor of the activation in the corresponding voxel of the target brain. The strength of this across-subject correlation measure stems from the fact that it allows the detection of all sensory-driven cortical areas without the need of any prior assumptions as to their exact functional role.

relation during the movie (Fig. 2). Interestingly, the correlation extended far beyond the visual and auditory cortices, to the entire superior temporal (STS) and lateral (LS) sulcui, retrosplenial gyrus and even to secondary somatosensory regions in the post-central sulcus, as well as multi-modal areas in the inferior frontal gyrus, and parts of the limbic system in the cingulate gyrus. The level of inter-subject correlation appears to provide a new quantitative 'social' measure for the stereotypical involvement of cortical areas with external sensory stimuli.

In order to rule out the possibility that the across subject correlations were introduced by scanner noise or preprocessing procedures, we measured the inter-subject

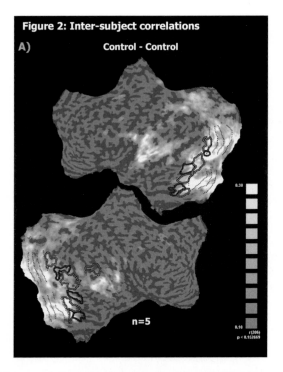

FIG. 2.   Intersubject correlation within the control subjects. (A) The average intersubject correlation across all pairwise comparisons within the healthy subjects. Correlation maps are shown on unfolded left and right hemispheres. Shade indicates the significance level of the inter-subject correlation in each voxel. Black dotted lines denote borders of retinotopic visual areas V1, V2, V3, VP, V3A, V4/V8. The face, object and building-related borders (dark grey, black and light grey rings respectively) are also superimposed on the map. Note the substantial extent of intersubject correlations, and the extension of the correlations beyond visual and auditory cortices.

correlations between brains, while subjects were placed in complete darkness. As opposed to the movie, during darkness there was almost no inter-subject correlation (2% ± 3 SD). Thus, the cause for the strong intersubject correlation is likely to be the engaging power of the movie to evoke a remarkably similar activation across all viewers.

What could be the specific source of the robust inter-subject correlation we found? In past research, developmental psychologists (Vygotsky 1962), social psychologists (Dijksterhuis & Bargh 2001), as well as philosophers (Wittgenstein 1951) stressed the central role which the external environment plays in shaping our thoughts, intentions and behaviours under different circumstances. This idea is nicely compatible with our recent finding of strong intersubject correlations between subjects imposed by the external movie stimuli. Thus, the tendency of individual brains to 'tick collectively' during natural viewing of the same movie, attests to the commonality of our neurobehavioural responses under similar external contexts.

However, in addition to the highly synchronized cortex, we also found a pattern of areas, which consistently *failed* to show inter-subject coherence. These areas included the supramarginal gyrus, angular gyrus and prefrontal areas. Thus, the 'collective' coherence measure naturally divides the cortex into a system of areas that manifest a stereotypical, intersubject response to external world stimuli, vs. regions that appeared to be linked to unique, individual variations. This heterogeneity hints at the possibility that the inter-subject correlations measure may be used as a new tool for tracing individually-unique cultural and attentional differences among various populations.

A particularly interesting possibility may be that, using the intersubject correlation measurement we could look for brain regions that exhibit social dysfunction within different clinical populations, such as, for example, autistic individuals. The autism disorder is characterized by difficulties in communication with the surroundings, leaving the autistic patients unable to interact with others and behaving compulsively and in rigid patterns. They also respond inappropriately to external stimuli and cannot generalize across environments. In collaboration with Dr. Marlene Behrmann from Carnegie Mellon, Pittsburgh, we compared the inter-subject correlation of the BOLD signal in a group of six autistic individuals to that of a control group all watching an identical popular movie. Our initial results reveal a strong and widespread disruption of the inter-subject correlation of the BOLD signal both within the autistic group and between it and the control group. The disruption of the correlation was not uniform but rather, was more evident as one moved from early visual and auditory sensory areas to higher-order association cortices. Thus, the use of intersubject correlation measures provides a new tool for tracking the deficits in neural response of autistic individuals to the external world under more naturalistic circumstances.

Finally, the intersubject correlation measurement can be used to search for inter-cultural differences in the interpretation of external stimuli. When inhabitants of the grate Amazon river region in south America were introduced to cinema in the early 1950s, they observed small details in the movie scenes (such as the chicken in the background) that were unnoticeable for Western audiences, while taking no notice of major events in the scene (such as the main plot). Nowadays, when TV and movies are part of all cultures, and differences between cultures diminish, it will be interesting to test whether the intersubject correlation measurement could be used for tracing inter-population differences within a community (e.g. children, teenagers and elderly people), as well as between cultures.

The reverse correlation

In order to identify the source of such powerful common 'consensus' between different individuals watching the same movie, we adopted an analysis approach loosely analogous to the 'reverse correlation' method used for single-unit mapping (Ringach et al 2002, Jones & Palmer 1987). In this analysis, we used the peaks of activation in a given region's time-course to recover the stimulus events that evoked them. Thus, constructing a regionally specific 'movie' which was based on the appended sequence of all frames which evoked strong activation in a particular region of interest (ROI), while skipping all weakly activating time points.

Using the reverse correlation methods we identified two unrelated components which were the main sources for the intersubject correlation: a spatially non-selective component and a regionally selective one. Below we discuss each component separately.

Non-selective activation component

Within each subject we identify a widespread, spatially non-selective activation wave that was apparent across many cortical areas. Importantly, this spatial non-selective activation wave was strongly correlated across different subjects during the movie (Fig. 3). Applying the reverse correlation to this global non-selective time course (black line in Fig. 3) revealed that it selected a substantial component of emotionally charged and surprising moments in the original movie (e.g. *all* gunshots and explosion scenes, or surprising shifts in the movie plot).

The spatially non-selective inter-area response could result from several factors. First, it could reflect modulations in the feed-forward processing load imposed by variations in the visual and contextual complexity of the movie scenes. Such global variations are expected particularly if the object representations have a widely distributed nature (Haxby et al 2001). Second, the non-selective component might reflect the global attentional and arousal impact of the scenes—as indeed was sug-

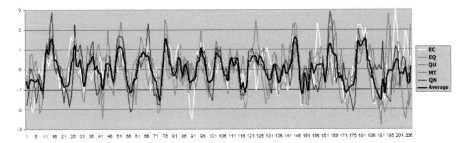

FIG. 3.   Correlated activation wave across five control subjects. The average non-selective time course across all activated regions, obtained during the movie for all five control subjects. The black line represents the across subject average time course. Note the striking degree of synchronization among different individuals watching the same movie.

gested by the highly emotional activation 'clips' generated through reverse correlation of the non-selective component with the movie. Although such arousal effects might also produce global autonomic responses, this is unlikely to explain the results, which show a highly stereotyped and heterogeneous neuroanatomical distribution (see Fig. 2).

Finally, it is interesting to note that in autistic individuals the spatially nonselective activation was much more widespread than in control subjects, but was not correlated across the autistic individuals.

Selective activation component

In addition to the global non-selective activation, the movie evoked distinct activation patterns in different brain regions—which were nevertheless highly correlated *between* individuals watching the same movie. This was assessed by performing the same across-subject correlation to the movie data set after the removal of the nonselective component from each voxel's time course. The reverse correlation method was able to recover known functional selectivity of cortical areas as of the face-related posterior fusiform gyrus (pFs; also termed the FFA [Kanwisher et al 1997, Haxby et al 2000]) and the building-related collateral sulcus (CoS; also termed the PPA [Aguirre et al 1998, Epstein & Kanwisher 1998]). These results are also compatible with those of Bartels & Zeki (2004a). In addition, the reverse correlation has proven to be very effective in uncovering unexpected functional specializations. Thus, the approach could serve as a promising unbiased tool for probing functional characteristics of new brain areas. This was most evident in the unexpected finding that the hand-related somatosensory region in the post-central sulcus was activated by images of delicate hand movements. It appears likely that this activation is part of the visuo–somatomotor 'mirror' system originally reported in the macaque monkey (Rizzolatti et al 1996, 1997) and more recently extended to the human cortex (Cochin et al 1999). Recently, social psychologists have stressed the role of such a mirror-system, directly linking perception and behaviour, as one of the fundamental bases of social cognition (Dijksterhuis & Bargh 2001).

Given the complete lack of control over the stimuli, the fact that we could recover the known functional selectivity of cortical areas during the movie watching has important implications concerning the nature of object representations. These become apparent when considering which constraints have been relaxed during the movie. First, subjects' eye movements were completely uncontrolled, which allowed subjects to view the stimuli at different retinal locations. The consistency of the results despite the spontaneous nature of eye movement is compatible with the notion that high order object areas are not very sensitive to changes in retinotopic position (Grill-Spector et al 1999, Ito et al 1995).

Second, the finding that the selectivity did not decrease with the movie's spatiotemporal complexity, points to the efficient operation of selection mechanisms, which govern subjects' attention (Brefczynski & DeYoe 1999, O'Craven et al 1999, Tootell et al 1998). Thus, given that several objects were embedded within a complex background in each scene, it seems that object-based attention (O'Craven et al 1999, Sheinberg & Logothetis 2001) was capable of isolating the object of choice within the complex frame.

Limitations

The use of more realistic and unbound experimental conditions, in combination with the analysis methods introduced above, provides new opportunities for novel experimental protocols and data analysis methods that embrace the complexity of natural stimulation and behaviour. However, as in any approach, these methods also have their limitations.

First, given that the current functional magnetic resonance imaging (fMRI) spatial resolution is in the order of millimetres the intersubject correlation can reveal only gross similarities between different fellow brains. Even so, given that the reverse correlation method has proven to be very effective in uncovering both known and unexpected functional specializations, it appears that the inter-subject correlation is sensitive enough to disclose refined functional discrimination which can be detected by using more conventional fMRI methods. Moreover, given that the alignment between different brains was based on the crude volume based Talairach coordinate system, we believe that our estimate of the inter-subject correlations, at the fMRI resolution, is still conservative, and could be improved by using more advanced cortex based alignment methods. Finally, given that the neuronal resolution of the functional columns is in the order of hundreds of microns it will be interesting to test at what level of organization these correlations are breaking apart.

Second, the rather sluggish nature of the haemodynamic response can make this method unsuitable when presenting a particularly rapid movie sequence (as in MTV video clips). However, as opposed to rapidly changing sequences, it seems that the natural temporal flow of events in real life situations tends to be rather slow, and sufficient for BOLD resolution. Thus, although no 'blank' periods were introduced to segregate BOLD responses in the present study, our results show that the BOLD signal was able to 'pick' quite successfully the object-selective frames appropriate for each cortical area under free and natural viewing conditions.

Third, given the complexity and multidimensionality of each frame (audio-visual contextual events) it will obviously be impossible to isolate the appropriate functional dimensions based solely on the reverse correlation methods. Thus, the reverse correlation should be viewed as a tool for 'pilot' search, both for normal and pathological cases, which can suggest preliminary functional specializations to be followed by a more controlled set of stimulation conditions.

Conclusion

Major trends in neuroscience research follow a reductionistic, deductive line of reasoning. For decades, neuroscientists have worked toward increased behavioural control and simplification, using precisely parameterized stimuli and behavioural tasks in highly controlled laboratory settings. This approach has obvious advantages and has served us well, as evidenced by the tremendous amount of knowledge amassed about brain structure and function. These conditions, however, are removed from natural real life situations. Thus, the *natural viewing* fMRI paradigms discussed here, which include the inter-subject correlation and the reverse correlation analysis, nicely complement the more conventional approaches by opening numerous novel possibilities for studying the neuronal correlates of complex human behaviours (such as social cognition, memory under real life situations, intercultural differences, etc.) under more natural settings.

Acknowledgements

Supported by the Human Frontier Science Program (HFSP), the Horowitz Foundation and the Benozio Center for Neurodegenerative Diseases.

References

Aguirre GK, Zarahn E, D'Esposito M 1998 An area within human ventral cortex sensitive to 'building' stimuli: evidence and implications. Neuron 21:373–383

Bartels A, Zeki S 2004a Functional brain mapping during free viewing of natural scenes. Hum Brain Mapp 21:75–85

Bartels A, Zeki S 2004b The chronoarchitecture of the human brain–natural viewing conditions reveal a time-based anatomy of the brain. Neuroimage 22:419–433

Brefczynski JA, DeYoe EA 1999 A physiological correlate of the 'spotlight' of visual attention. Nat Neurosci 2:370–374

Cochin S, Barthelemy C, Roux S, Martineau J 1999 Observation and execution of movement: similarities demonstrated by quantified electroencephalography. Eur J Neurosci 11:1839–1842

Dijksterhuis A, Bargh JA 2001 The perception-behavior expressway: automatic effects of social perception on social behavior. In: Zanna MP (ed) Advances in experimental social psychology. Academic Press, San Diego, p 1–40

Epstein R, Kanwisher N 1998 A cortical representation of the local visual environment. Nature 392:598–601

Grill-Spector K 2003 The functional organization of the ventral visual pathway and its relationship to object recognition. In: Kanwisher N, Duncan J (eds) Attention and performance XX: Functional brain imaging of visual cognition. Oxford University Press, Oxford, p 169–194

Grill-Spector K, Kushnir T, Edelman S et al 1999 Differential processing of objects under various viewing conditions in the human lateral occipital complex. Neuron 24:187–203

Hasson U, Harel M, Levy I, Malach R 2003 Large-scale mirror-symmetry organization of human occipito-temporal object areas. Neuron 37:1027–1041

Hasson U, Nir Y, Levy I, Fuhrmann G, Malach R 2004 Intersubject synchronization of cortical activity during natural vision. Science 303:1634–1640

Haxby JV, Hoffman EA, Gobbini MI 2000 The distributed human neural system for face perception. Trends Cogn Sci 4:223–233

Haxby JV, Gobbini MI, Furey ML et al 2001 Distributed and overlapping representations of faces and objects in ventral temporal cortex. Science 293:2425–2430

Ishai A, Ungerleider LG, Martin A, Schouten HL, Haxby JV 1999 Distributed representation of objects in the human ventral visual pathway. Proc Natl Acad Sci USA 96:9379–9384

Ito M, Tamura H, Fujita I, Tanaka K 1995 Size and position invariance of neuronal responses in monkey inferotemporal cortex. J Neurophysiol 73:218–226

Jones JP, Palmer LA 1987 The two-dimensional spatial structure of simple receptive fields in cat striate cortex. J Neurophysiol 58:1187–1211

Kanwisher N, McDermott J, Chun MM 1997 The fusiform face area: a module in human extrastriate cortex specialized for face perception. J Neurosci 17:4302–4311

O'Craven KM, Downing PE, Kanwisher N 1999 fMRI evidence for objects as the units of attentional selection. Nature 401:584–587

Ringach DL, Hawken MJ, Shapley R 2002 Receptive field structure of neurons in monkey primary visual cortex revealed by stimulation with natural image sequences. J Vis 2:12–24

Rizzolatti G, Fogassi L, Gallese V 1997 Parietal cortex: from sight to action. Curr Opin Neurobiol 7:562–567

Rizzolatti G, Fadiga L, Gallese V, Fogassi L 1996 Premotor cortex and the recognition of motor actions. Brain Res Cogn Brain Res 3:131–141

Sheinberg DL, Logothetis NK 2001 Noticing familiar objects in real world scenes: the role of temporal cortical neurons in natural vision. J Neurosci 21:1340–1350

Tootell RB, Hadjikhani N, Hall EK et al 1998 The retinotopy of visual spatial attention. Neuron 21:1409–1422

Vygotsky LS 1962 Thought and language. MIT Press, Cambridge, MA

Wittgenstein L 1951 Philosophical investigations. Prentice Hall, NJ

DISCUSSION

Logothetis: If you had a large enough magnet to put a symphony orchestra in, you would probably record the most unbelievable intersubject correlations. If one of the violin players were an idiot, you would also see a lack of correlation. But would this be useful at all? What is it telling us?

Hasson: Yes, revealing consistent correlations as well as mismatches between groups of subjects can be very useful. We all know that using simple and precisely parameterized stimuli in highly controlled laboratory settings cannot capture the bulk of our cognitive capacity. Therefore, the idea of looking at the functional proprieties of the human brain in real life situations is highly challenging although extremely difficult. The use of the spatiotemporal activity patterns in one brain to 'model' activity in other fellow brains gives us a new tool for tracing systematic differences in how different experimental groups respond in real life situations. For example, we can use the inter-subject correlation for tracing differences as well as similarities in the way men and women perceive the world, or to search for such differences across cultures or across clinical groups. Another promising avenue is to look for individual differences rather then between groups differences. Usually, and rightly so, most of our studies are hypothesis driven. However, our data-driven intersubject correlation approach can provide us with opportunities for novel experimental protocols and data analysis methods that embrace the complexity of natural stimulation and behaviour.

Logothetis: I hope you can do much more than you have presented here. If we put it in simple terms, everyone has a motor cortex and everyone's motor cortex is going to give you some kind of similar pattern if you have the same movements. When they watch a movie, almost all people will see the same things unless they have perceptual problems, and the signals they give will look similar. I fail to see how we can learn about the brain by doing these sorts of studies.

Hasson: I think that Nikos is raising an important concern regarding the intersubject correlation; that is, how sensitive it is to refined differences across subjects. Or in other words, could the intersubject correlation reveal only crude correlations across subjects (e.g. activation of Broca's Area during language processing) or could it reveal more subtle differences (e.g. refined grammatical transformations within this area)? This is still an open question waiting for further investigation, but our preliminary findings suggest that the intersubject correlation has proven to be very effective in uncovering both known refined functional selectivity (as the selective activation in the PPA and FFA) as well as unexpected new functional specializations (as the activation to hand actions in the post central sulcus).

Moreover, currently the inter-subject correlation is based on the crude volume based Talairach coordinate system alignment. So, there are ample opportunities to increase the sensitivity of our methods by improving our alignment techniques.

Dehaene: I found it interesting that there were many areas that were not synchronized across subjects. I am struck, in particular, that the parietal–frontal–cingulate system seemed to be completely quiet and didn't show any intersubject correlations. Can you distinguish between two possible explanations of this observation? One explanation is that you simply didn't engage these systems, because you did not provide any stimulus that would have been relevant for them. However, another, more interesting explanation might be that there are some areas that, because of their spontaneous level of activity, are actually very active but in a manner which is autonomous and detached from the outside world, and therefore not synchronized among subjects. These areas would have a high-level of spontaneous activity and autonomy relative to outside stimuli.

Hasson: I agree. The fact the there are brain areas in each subject that respond strongly during the movie watching but nevertheless their response profile is uncorrelated across subjects might enable us to capture true individual differences across different viewers. Of course there could be many sources of variability so we should be really cautious when trying to take apart true variability from noise. We were thinking of two ways to approach this intriguing question. First, by showing the same movie twice to each subject and computing the correlation between the two presentations within each subject, we can look for brain areas that are correlated within a given subject but not between subjects. Such areas could reveal some functional processes that are unique to a given individual. Second, we can try and correlate each unique signal with an independent behavioural measurement (e.g. the level of anxiety in each individual) in an attempt to account for the variance across individuals.

Rizzolatti: My first feeling when I read your article in *Science* was rather similar to that just expressed by Nikos Logothetis. Very nice paper, but is it really important? Now, thinking about the possible application of the technique described in that paper to different groups of volunteers and especially to patients with specific disturbance, I think your contribution is really valuable.

Logothetis: If you take a movie like that and show it to the same person 10 times, I would bet my pension that you would get half of these patterns from an anaesthetized person. The connectivity is so strong. There is no way to establish identical conditions for each of the presentations of the movie, particularly if it is between subjects where there will be large differences in motivation, how much they like it, attentional aspects and so on. This will mask things that are subtle. You will see just robust activations.

Hasson: You are worried that the movie provides too rich and uncontrolled stimuli to result in any interesting correlations. I had the same worry when I started using such rich and complex natural stimuli. However, our results argue for the opposite. In fact, we did not see an increase in the variability across subjects during the movie watching relative to controlled experiment. Even more so in some cases we show that the response to a single event in the movie was in remarkable agreement across all subjects. One of the main reasons for such a strong correlation across subjects during the movie watching is that our brains are evolved to cope with such complex and dynamic stimuli. After all, movie watching is part of our daily life experience. Moreover, our results attest to the engaging power of the movie to evoke a remarkable activation across all of us. In a sense a good movie could be conceived of as the ultimate edited experiment which succeeds to manipulate our perception by carefully stimulating all our senses.

Logothetis: I have nothing against the stimulus. The stimulus is an excellent one. We use movie scenes with monkeys; they are incredibly rich. If you use a movie scene you will get a large activation everywhere. I just wonder whether there is anything interesting in comparing subjects in this way.

Diamond: I'd like to concentrate on the face area. Your experiments showed that in all subjects this area is activated when they looked at faces. Besides knowing that we are seeing a face, the brain also wants to identify the face among thousands of possible alternative faces. Is there any prospect that your kind of method could show how the neurons in the region are distinguishing one face from another?

Hasson: It is an interesting question. Our results clearly show that during the movie the FFA area was strongly activated by close-up of faces. This result reveals that the fusiform selectivity to faces, reported by now in so many controlled experiments, is actually preserved under natural and unbound settings. Nevertheless the role of the FFA in face recognition is still under debate, and I am afraid that we will need further experiments to study this issue. For example recently we scanned a congenital prosopagnosic subject, and discovered that he has a normal face selective activation in the fusiform gyrus. It will be interesting to test how such congenital subjects respond under such natural and unbound viewing conditions.

Krubitzer: I was interested by the overall activation of the entire brain during fearful scenes. The brain is a really expensive machine to run. Under normal conditions are we sitting here listening to the talks with some set level of activation? Nikos Logothetis drops a bomb by suggesting all this is trivial and we suddenly change general cortex state. Do you think this is reasonable?

Hasson: The role of the global elevation in signal in particular moments during the movie is still not clear. The reverse correlation analysis indicates that this might be related, at least partly to arousal and attention factors. So your suggestion that the common signal might be an efficient way of the brain to allocate resources to significant events in the external world is highly feasible.

Krubitzer: It may just be a way to conserve energy.

Harris: It could be generally a result of an increase in blood flow.

Krubitzer: Yes, but it is hard to separate these out.

Romo: Was this in the whole brain or just a few sensory areas?

Hasson: It wasn't across the entire brain, but was confined to regions that were activated during the movie watching. For example when we removed the sound track from the experiment this elevation was not seen any more in the auditory cortex.

Brecht: Even if these real life experiments come up with trivial results, they justify our stupid experiments. But they can also produce non trivial results. If Uri Hasson didn't find what we predicted, then this would be much more worrying.

Porro: Have you tried to relate the general changes in average signal across the cortex with changes of arousal?

Hasson: We are currently trying to correlate the cortical changes with skin conductance.

Representation of object images by combinations of visual features in the macaque inferior temporal cortex

Manabu Tanifuji*, Kazushige Tsunoda*[1] and Yukako Yamane*[2]

*Laboratory for Integrative Neural Systems, Brain Science Institute, The Institute of Physical and Chemical Research (RIKEN), Hirosawa 2-1, Wako-shi, Saitama 351-0198, Japan

Abstract. The ventral visual pathway is essential for object recognition where features necessary for recognition are extracted from object images. This pathway is not directly related to action; nevertheless extraction of features from object images through this pathway is closely related to actions since we determine our behaviour based on sensory information, such as object images in the scene. To link perception and action, we have investigated neural representation of object images in area TE. Area TE, the anterior part of the inferior temporal (IT) cortex, is the final purely visual area in the ventral visual pathway. Since individual neurons in this area respond to features less complex than object images, an object image is supposed to be represented by multiple features. Here, to investigate sets of features necessary for representation of object images, we conducted a combination study of an optical imaging technique and single cellular recordings from anesthetized macaque monkeys.

2005 Percept, decision, action: bridging the gaps. Wiley, Chichester (Novartis Foundation Symposium 270) p 217–231

Visual features represented by neurons in area TE

It is known that neurons in this area respond to object images but also to visual features that are geometrically less complex than the object images (Desimone et al 1984, Tanaka et al 1991, Kobatake & Tanaka 1994). In particular, Tanaka and colleagues examined responses of neurons in area TE with systematically simplified visual stimuli and found that visual stimuli sufficient to activate many neurons were moderately complex visual features that are less complex than object images ('critical feature') (Kobatake & Tanaka 1994) (e.g. see Fig. 1). Many of these features are

[1]Current address: Laboratory for Visual Physiology, National Tokyo Medical Center, Higashigaoka 2-5-1, Meguro-ku, Tokyo 152-8902, Japan.
[2]Current address: Krieger Mind/Brain Institute, The Johns Hopkins Univ. 3400 N. Charles St., Baltimore, MD 21218, USA.

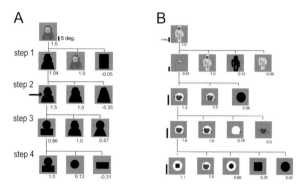

FIG. 1. The 'critical feature', a visual feature that maximally activates each cell, is determined by systematic stimulus simplification of the best object stimulus. First, we tested the cell with various 3D objects including faces, hands, stuffed animals, plastic fruits and vegetables, and paper mounts. After determining the best stimulus, we simplified it step by step to find the simplest stimulus that maximally activates the cell. For example in (A), at step 1, we compared the best coloured object with its silhouette, and found that the silhouette activated the cell equally well. The rightmost rectangle was taken as a control stimulus. The numbers below each picture indicate the response amplitudes normalized to the response to the reference stimulus, the best object. At step 2, we examined the effect of the 'sharpness' of the corner at the junction of upper and lower parts (arrow), and found that the silhouette with the sharp corners was the most effective stimulus. From left to right, the stimuli were the silhouette with sharp corners, the silhouette that evoked the best response at the previous step, the silhouette without corners. Further simplification was carried out at step 3. The leftmost stimulus evoked more than 70% of the response elicited by the best stimulus in the previous step, and was again examined in the next step as the reference stimulus. Finally at step 4, we determined the critical feature as a combination of a circle and a rectangle because neither the upper nor lower part alone activated the cell. Panel (B) shows another example of stimulus simplification where we found a dark square within a white circle as the critical features. (Y. Yamane and M. Tanifuji, unpublished observations.)

combinations of simple shapes, colours, luminance gradient/contrast, and textures. Although these features are more complex than the optimal visual stimuli for cells in areas V1, V2 and V4, they are still less complex than natural objects (Fig. 1).

Optical imaging to investigate distributed activity in area TE

Fujita and his colleagues showed that neurons with similar response properties are clustered into a column in area TE (Fujita et al 1992). Thus, intrinsic signal imaging of columnar activation can be used to investigate spatial patterns of activation (Tsunoda et al 2001, Wang et al 1996, 1998). This technique measures the decrease in the degree of light reflection elicited by neural activation from the exposed cortical surface using a CCD camera; the reflection changes are due to metabolic changes elicited by neural activation including changes in deoxygenation of haemoglobin in capillaries (Grinvald et al 1999). Intrinsic signal imaging in area TE

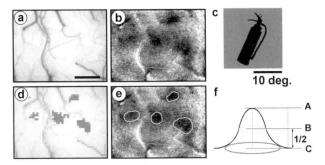

FIG. 2.   Intrinsic signal imaging detects local modulation of light absorption changes in area
TE. (a) Portion of area TE where intrinsic signals were recorded. (b) A differential image showing
a local increase in absorption by the visual stimulus shown in (c). (d) Active spots, where the
degree of reflection change evoked by the stimulus was significantly greater than that without
the stimulus presentation. The region with the highest significance level is in black, and that with
the lowest significant level in white ($P < 0.05$). (e) Active spots outlined by connecting pixels with
1/2 of the peak absorption value as in (f). (Modified from Tsunoda et al 2001.)

revealed that visual stimulus elicits local decrease of light reflection that appears as
spots distributed across the cortex surface (Fig. 2) (Tsunoda et al 2001). Although
these reflection changes are not a direct measure of neural activation, intrinsic
signals coincide well with the activity of neurons examined by conventional extracellular recordings (Wang et al 1996, 1998, Tsunoda et al 2001). These spots, 'active
spots', could correspond to a column of cells with similar responsiveness in this
area (Fujita et al 1992).

Neural representations of object images in area TE

Intrinsic signal imaging with various object stimuli revealed that complex objects
activate multiple spots (Fig. 3A). Some of the spots were commonly activated by
different objects; other spots were specific for one of the examined stimulus. This
observation is consistent with the idea that each of these spots could represent a
particular visual feature as proposed previously. To examine this idea, we compared
distribution patterns of spots activated by a complex object with those activated by
systematically simplified stimuli (Fig. 3B) (Tsunoda et al 2001). We used a 'black cat'
(a-*1*) as the complex object image and then simplified it to its 'head' (a-*2*), and to
the 'silhouette of its head' (a-*3*). The original image (a-*1*) elicited fourteen spots,
but presenting the 'head' (a-*2*) elicited only eight of the original 14 spots. The silhouette (a-*3*) only activated three (white) of the eight spots elicited by the head
(a-*2*). Thus, the simplified stimulus lacking part of features in the original image
activates only a subset of the spots elicited by the stimuli before simplification.
Thus, individual spots represent visual features rather than object images in area

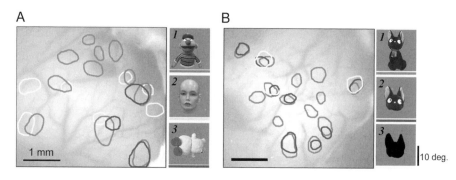

FIG. 3. Spatial patterns of activation elicited by complex object images are mapped on the image of the exposed cortical surface where the optical signals were recorded. (A) Activity in spots was elicited by three different object images. The colour of outlined spots and the line underneath the stimulus image are matched for the stimulus and the activated spots. (B) A case where simplified stimuli elicited only a subset of spots evoked by more complex stimuli.

TE. The actual visual features represented by individual spots were examined by extracellular recordings from neurons in active spots. For example, two spots (A, B) were activated by images of a fire extinguisher and a silhouette of it, but not by the fire extinguisher without handle and hose (Fig. 4). This difference in optical response patterns suggests that spots A and B represented visual features related to the handle and hose of the fire extinguisher. To confirm this expectation, we recorded extracellular responses from 25 cells in these spots shown in Figure 4, and analysed the response properties of the cells in each spot (Fig. 5). The handle and hose in isolation as well as the silhouette of the original fire extinguisher activated cells in spots A and B. The cells in spot A were also activated only by the handle having protrusions, but not by the hose alone. Furthermore, other stimuli with sharp protrusions, such as a 'hand' and 'cat's head' also activated the cells. Thus, the critical feature for the cells in spot A was 'sharp protrusions'. In contrast, cells in spot B were activated by the hose alone, but neither by the handle nor a 'line segment'. Thus, the critical feature for the cells in spot B was an 'asymmetric arc'. The cells in spot C were significantly activated by both the original fire extinguisher and the one without handle and hose. The simplest visual feature for cells in spot C was a 'rectangular shape', but cells also responded significantly to an 'ellipse'. Since there was no response to a 'circle', we determined the critical feature of the spots as an 'elongated structure'.

Representation of spatial arrangement of parts in object images

Examination of visual features represented by neurons in area TE suggested that at least some of the neurons in this area, represent 'local features' in object images,

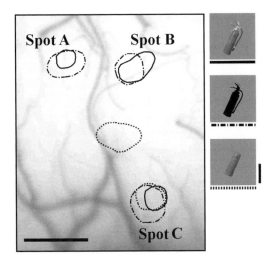

Spot A   Spot B

Spot C

FIG. 4. A case shows that some spots (A and B) were activated by a fire extinguisher but not by the one without handle and hose. The spot at the center showed an opposite response pattern: activation with the fire extinguisher without handle but not by the original one. See Tsunoda et al (2001) for detailed analysis of the spot at the centre. Horizontal scale bars, 1 mm; vertical scale bar, 10 degrees.

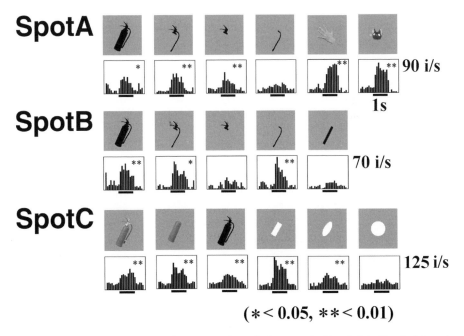

SpotA   90 i/s
                                                    1s

SpotB   70 i/s

SpotC   125 i/s

(∗ < 0.05, ∗∗ < 0.01)

FIG. 5. Visual responsiveness of representative cells in spots A, B and C in Fig. 4. Adapted from Tsunoda et al (2001).

as neurons in spots A and B (Fig. 4) represent 'protrusions' and 'asymmetric curvature', respectively. Thus, information about the spatial arrangement of 'local features' is necessary for the specific representation of object images; some of the other spots may represent visual features related to the spatial arrangement of local features ('configurational information'). Here, we refer to 'local features' as visual features that occupy part of an object image and are distinguishable from other parts of an object image by their particular shapes, colours, or textures. 'Configurational information' is information about the spatial relationship of 'local features' themselves or about the spatial relationship of parts including local features. To examine the representation of 'configurational information', we investigated spots activated by an object (Original, Fig. 6-1) and the same object with a gap introduced between parts of the object (Fig. 6-4), but not by a part alone (Figs. 6-2 and 6-3) (Yamane et al 2001). These spots do not simply represent local features in objects because either part is not essential for activation. Moreover, activation by the stimulus with an introduced gap indicates that local features appearing at the junction of two parts, such as sharp connecting corners in Fig. 6-1, are also not essential. In three monkeys, we indeed found some of active spots had stimulus selectivity as described above. Extracellular recordings from cells in these spots showed that their critical features were combinations of vertically aligned two parts (Fig. 7a). In particular, the stimulus simplification procedure for these cells in this spot revealed that there was no activation by either part (for a representative case, see Fig. 1A). These cells were less sensitive to colour, texture, and local shapes of either part:

- there were no changes in the responses after removing colour and texture during the stimulus simplification procedure (Fig. 1A and 7a), and
- these cells responded equally well to object images even with different colour, texture, and local shapes, as long as the global configuration was similar to the critical features (Figs. 7b).

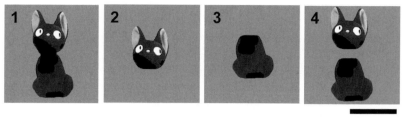

10 deg.

FIG. 6.   A representative set of visual stimuli used in intrinsic signal imaging for examination of the representation of the spatial arrangement of parts. We searched for the spots that are activated by stimuli 1 and 4, but not by stimuli 2 and 3.

FIG. 7. Effective stimuli for neurons in a spot identified by the stimuli in Fig. 6. (a) Representative critical features determined by the stimulus simplification (Fig. 1). Please note that when colour and texture are not essential, the stimulus was filled black (see Fig. 1A). (b) The best object stimuli for these neurons, among 100 object stimuli examined before stimulus simplification. Scale bar, 5 degrees.

In contrast, we found that these cells were highly selective to a particular spatial arrangement of the upper and lower parts (Fig. 8).

Summary and conclusions

The combination of intrinsic signal imaging and extracellular recordings suggests that object images are represented as combinations of spots, and that each spot represents visual features less complex than the original object images (Figs. 4 and 5). These visual features are not specific for a certain object image but are common features across the wide range of object images. For example, spot A in Figure 4 represents 'sharp protrusions' that is a common feature among a fire extinguisher, hand and cat (Fig. 5 spot A). Thus, representation specific to object images requires combination of spots.

Spots do not necessarily represent 'local features' but some of them represent visual features related to object configurations. We found that neurons in these spots responded to visual stimuli consisting of vertically aligned upper and lower parts

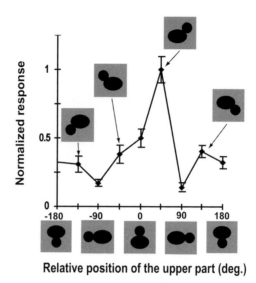

Relative position of the upper part (deg.)

FIG. 8. Response specificity of a representative cell to the spatial arrangement of parts. The upper part of the critical feature of the cell was rotated relative to the lower part. The horizontal axis indicates the angle between the line connecting the center of upper and lower parts of each stimulus and that of the critical feature. The vertical axis indicates normalized value of stimulus evoked responses. In this particular case, the best response was elicited by the stimulus with 45 degrees, but many others respond maximally at 0 degrees.

(Fig. 8), but were less selective to local features embedded in either part (Figs. 1A and 7). We consider that such neurons specify the configuration of object images. Face neurons in area TE could play the same role. They respond specifically to a configuration specific for faces, but are less selective to individual faces (Desimone et al 1984, Perrett et al 1984, Baylis et al 1985, Yamane et al 1988, Young & Yamane 1992).

Although we made a distinction between local features and the spatial arrangement of parts in this text for clarity, we do not necessarily consider that neurons in area TE are generally categorized in two groups which represent either local features or spatial arrangement of parts. It is known that visual responses of neurons in inferior temporal cortex are modified by visual experience (Miyashita 1988, Logothetis et al 1995, Kobatake et al 1998), and we consider that visual features represented by neurons are shaped based on the statistical nature of the visual environment. A single neuron could represent partly global and partly local features if such a combination appears frequently in the visual environment. We observed neurons representing local features such as protrusion and curvature, those representing vertically aligned two parts, and those responding to faces, because these features are common in the natural visual environment.

Finally, in relation to action, our present knowledge about representation of object images in area TE may not be sufficient. Although our investigations suggest that multiple spots representing visual features of an object image are necessary to specifically represent the object image, we may not need complete representations of object images in a behavioural context. Investigations described here with behaving animals are required in future.

Acknowledgments

This work was partly supported by Research Fellowships of the Japan Society for the Promotion of Young Scientists to Y. Y.

References

Baylis GC, Rolls ET, Leonard CM 1985 Selectivity between faces in the responses of a population of neurons in the cortex in the superior temporal sulcus of the monkey. Brain Res 342:91–102

Desimone R, Albright TD, Gross CG, Bruce C 1984 Stimulus-selective properties of inferior temporal neurons in the macaque. J Neurosci 4:2051–2062

Fujita I, Tanaka K, Ito M, Cheng K 1992 Columns for visual features of objects in monkey inferotemporal cortex. Nature 360:343–346

Grinvald A, Shoham D, Shmuel A et al 1999 In-vivo optical imaging of cortical architecture and dynamics. In: Windhorst U, Johansson H (eds) Modern techniques in neuroscience research. Springer, Berlin, p 893–970

Kobatake E, Tanaka K 1994 Neuronal selectivities to complex object features in the ventral visual pathway of the macaque cerebral cortex. J Neurophysiol 71:856–867

Kobatake E, Wang G, Tanaka K 1998 Effect of shape-discrimination training on the selectivity of inferotemporal cells in adult monkeys. J Neurophysiol 80:324–330

Miyashita Y 1988 Neuronal correlate of visual associative long-term memory in the primate temporal cortex. Nature 335:817–820

Logothetis NK, Pauls J, Poggio T 1995 Shape representation in the inferior temporal cortex of monkeys. Curr Biol 5:552–563

Perrett DI, Smith PAJ, Potter DD et al 1984 Neurons responsive to faces in the temporal cortex: studies of functional organization, sensitivity to identity and relation to perception. Hum Neurobiol 3:197–208

Tanaka K, Saito H, Fukada Y, Moriya M 1991 Coding visual images of objects in the inferotemporal cortex of the macaque monkey. J Neurophysiol 66:170–189

Tsunoda K, Yamane Y, Nishizaki M, Tanifuji M 2001 Complex objects are represented in macaque inferotemporal cortex by the combination of feature columns. Nat Neurosci 4:832–838

Wang G, Tanaka K, Tanifuji M 1996 Optical imaging of functional organization in the monkey inferotemporal cortex. Science 272:1665–1668

Wang G, Tanifuji M, Tanaka K 1998 Functional architecture in monkey inferotemporal cortex revealed by in vivo optical imaging. Neurosci Res 32:33–46

Yamane S, Kaji S, Kawano K 1988 What facial features activate face neurons in the inferotemporal cortex of the monkey? Exp Brain Res 73:209–214

Yamane Y, Tsunoda K, Matsumoto M, Phillips A, Tanifuji M 2001 Decomposition of object images by feature columns in macaque inferotemporal cortex. Soc Neurosci Abstr 27:1050

Young MP, Yamane S 1992 Sparse population coding of faces in the inferotemporal cortex. Science 256:1327–1331

DISCUSSION

Esteky: When you choose a spot in your optical imaging and stick the electrode in, do you select a neuron that is responsive to the stimulus you are using, or do you randomly select any neuron? In a given cortical column in inferior temporal cortex (IT) there are a variety of cells responsive to many different kinds of stimuli. So you are not dealing with a uniform and homogeneous cortical spot. Do you select the stimuli that fit your optical imaging results?

Tanifuji: No. We were only biased to neurons which gave visual responses. In other words, we didn't examine neurons that didn't reveal any visual responses. Here, visual responses were examined for about 100 3D objects. These objects were manually presented to animals. This procedure allowed us to show more than 100 visual stimuli, because each object was presented at different viewing angles.

Esteky: Was there always a good correspondence between the optical imaging data and the single unit data, or were there inconsistencies?

Tanifuji: It depends on spots. The correlation is very high in some spots where more than 90% of neurons show response consistent with the result of optical imaging. However, in other cases, the percentage is as low as 60%. But if we randomly select neurons regardless of the location of spots, the number of neurons responding to the stimuli that activated a certain spot is less than 30%. The value, 60% means that there is some clustering of neurons with similar response property, and thus we could see activated spots by optical imaging. I somehow wonder whether there could be strong and weak columns depending on how variable the response properties of neurons within the columns are.

Logothetis: I have a general question regarding work with different types of shape descriptions. I know how tedious it is to do this kind of work. Were you able to get a sense of what kind of variability and among what kinds of dimensions this variability is in single spots? If you have two or three presentations in approximately the same position could you say what kind of variability you found with this systematic approach?

Tanifuji: It is very difficult to explain in a few words.

Albright: I assume you are talking about variability across neurons. These things are 500 microns across. It can't possibly be the case that there are consistent properties within that entire region. In our experience you can go from one neuron to the next and find a completely different set of recordings.

Logothetis: So if you get the parameters of the first two coefficients of the Fourier descriptors you are using, you find huge discontinuity.

Albright: Yes, which is somewhat inconsistent with this notion of columns in IT. The columns are categorical, but within one of those columns there is a great deal of variability. We don't know what the relevant variables are.

Logothetis: In defining these columns, there is a process of categorization on the part of the experimenter. I see a certain computationally dangerous arbitrariness. There is not much of a systematic basis to what you see. One has a tendency to say there is a column there. What would be the definition of the column if we wanted to call these spots a column?

Tanifuji: Let me explain a simple experiment we did. We examined the correlation of selectivity for 100 objects for nearby neurons. Then, we found the correlation coefficient for the stimulus selectivity is 0.4. Though it is not very high, this value is still statistically significant.

Albright: Are the two neurons adjacent to one another in successive recordings?

Tanifuji: Yes. The pairs were placed near each other. The distance was approximately 150 μm.

Logothetis: Have you tried to do the best possible isolation of these neurons. Sometimes an 'isolated' neuron is 2 mV.

Tanifuji: We isolate the cell with the standard template matching. I think there is no particular bias to large or small cells. We analysed multiple cellular activities as well.

Logothetis: Have you tried to do the unit responses?

Tanifuji: Yes. I should point out one thing. Although a correlation coefficient of 0.4 is statistically significant, looking at the best object stimulus of individual cells, they are different. As reported by Ichiro Fujita and Keiji Tanaka (1992), however, the critical features of individual cells are very similar. We consider that although the object selectivity of nearby neurons is different, this is because the tuning broadness around the optimal stimulus is not the same for different neurons, and that the optimal stimuli, namely critical features, themselves are the same for nearby cells.

Logothetis: You said that there is a 0.4 correlation between the arbitrary tuning curves. This means that you only explain 16% of the variance. Would you call this a column? Tom Albright, do you remember what the situation was with MT?

Albright: It is better than 0.4. I want to make sure what the measurement is here. You have 100 stimuli and the correlation for responses across neurons is 0.4 for neurons within the same module. What is it if you have one within a module and another outside?

Tanifuji: Then, the correlation coefficient dropped down to around 0.1.

Esteky: On average, some 15% of stimulus images evoke significant responses in an IT cell. This suggests a rather wide shape tuning in IT. The wide tuning is true even for cells known to respond 'selectively' to faces: single unit recordings and fMRI studies have shown that face cortical cells/modules also respond to non-face stimuli.

Tanifuji: If neurons that respond to a particular stimulus are homogeneously distributed across the cortex, then optical imaging with the stimulus would not show any spots. Appearance of spots means that there is clustering of neurons at least for those that respond to that particular stimulus. One thing that is uncertain is how much darkening we could obtain if all the neurons within a spot are activated. Thus, even if darkening of a spot is observed with a stimulus, optical imaging experiments could not say anything about a possibility that some neurons in the spots did not respond to the stimulus but to others.

Esteky: Your approach does not reveal whether you are dealing with specific spots that respond exclusively to the selected stimulus or spots that are more widely tuned. In other words, in addition to showing that a stimulus is effective in activating a particular column or a cell (i.e. response selectivity) one needs to show how specific or exclusive the cells respond to that particular stimulus (i.e. response specificity). The correlation difference that you get in distant columns depends on the type of visual images used in your stimulus set. If the images contain a wide range of common or simple features a lower difference would be expected and vice versa.

Tanifuji: Of course different objects could activate the same cell. The question is whether this is because the cell is broadly tuned to these objects or the cell is tuned to a feature that is common to these objects, even though these objects look very different. In the case of the fire extinguisher, I showed a cell (spot A) that responded to the fire extinguisher, a hand, and the cat's head. These objects are different, but responses to these objects were easily explained by the cell's specific responses to 'protrusions'. We can not exclude the former possibility, but at least this result favours the latter possibility.

Let me say something more about the object selectivity. We compared object selectivity for nearby neurons and found that the correlation coefficient of stimulus selectivity for nearby cells is something like 0.4. This means that there is a statistically significant correlation in stimulus selectivity of cells, but it also means that the tuning curves of nearby cells are different to some extent. Then, one might imagine that the averaged activity of all the neurons in a column will lose stimulus specificity, or the tuning curve of averaged activity would not be as sharp as that of individual neurons. However, we found that this is not the case. We recorded many neurons in the same spot, and averaged neurons' activities for individual object to generate super multiple cellular activity. Next, we compared sharpness of object selectivity for super-multiple cellular activity, for single cellular responses, and for multiple cellular activities (MUA at a site). Interestingly, the sharpness of the object selectivity is almost the same for these three cases. Based on this observation, we think there is a certain feature that represents the spot, but this feature could not be simply captured by the result of object selectivity from one or two constituent neurons in the spot. The reason is that, in addition to common response property of the spot, individual cells have their own tuning variability, and that the variability in individual cells masks the common property of the spot in the object selectivity experiment. One way to capture the columnar level property is to use the reduction process used in previous studies by Keiji Tanaka's group (1991; 1992), they determined a critical feature of one cell and examined other cells along the vertical penetration with the stimuli related to the critical feature and found that the tuning property is similar for those cells. Their work was successful because they made proper selection of stimulus space to eliminate cell to cell variability.

Diamond: I have a technical question about the population tuning curve. If you were to move the electrode 500 μm and then do a new population tuning curve would it look different? The alternative is that there is some different saliency of the different objects, so the whole population is more responsive to some rather than others. That is, in a different penetration, some distance away, if you then get a population of neurons along the electrode track and for the whole population you show the same set of stimuli, is the population tuning curve different to the other spot?

Tanifuji: The tuning specificity of the other spot is different. The selectivity is totally different in a column that is 500 μm away: the correlation immediately dropped to 0.1 or lower. Putting it another way, to give you an idea how different columns are in their response properties: the selectivity of individual neuron is characterized as a point in the high dimensional stimulus space consisting of 100 axes each of which represents the evoked response of the cell to one of 100 objects. Then, since different cells in a spot have different selectivity, the points, each of which represent cells in a spot, are scattered around in this stimulus space. However, the region covered by these cells in this stimulus space is well (statistically significantly) separated from the region covered by the cells from the other spot. This result means that object selectivity is different from spot to spot, and that this difference exceeds the variability of the selectivity of cells within a column.

Treves: We are all asking versions of the same question, so let me ask my version, which has to do with the categorical nature of these putative columns. If you start with a large array of stimuli and perform optical imaging, you can define the spots for each stimulus in this large array. Then to what extent do you find spots that do not overlap precisely? This tells us something about whether there is any meaning to the boundaries of the columns, or whether the columns are clouds of neurons with some kind of shared selectivity. This is an important component of our notion of columns.

Tanifuji: If you show many objects to an animal and overlay all the spots elicited by these objects together, you may not see a clear boundary of spots. This is because the columns responding to similar features tend to partially overlap each other, such as orientation columns in V1. (In the analysis of object selectivity in high dimensional space, to make the point clear, we compared distant spots.) But if we use one object or derivative of it as stimulus, then because each object consists of combinations of different features, we can see activation of distant spots that change systematically by the manipulation of the stimulus.

Brecht: Do you have any idea how dark your spots are? This might also reveal something about the clustering. Is it as dark as the spot for orientation columns in V1? Or is it much weaker, or 10 times weaker?

Tanifuji: It is about 10 times weaker! But from this value we may not simply conclude that one over 10 neurons are activated by a stimulus, because evoked responses of individual neurons are also different from V1 cells.

Albright: The argument has been made that these cells contribute to object recognition. In as much as this is true, you would expect them to exhibit the sort of invariances that are characteristic of object recognition. Do you see these kinds of invariances if you keep the same local features but scale the thing up or rotate it, for example? Are the cells equally selective?

Tanifuji: We haven't tested object size in this study. In the literature people have showed a certain range of size invariance. We changed the spatial position. If we move the objects in space there is no change in evoked responses.

Albright: One of your stimuli was a ball on top of an ellipse, and you rotated the relative positions around. If this is done symmetrically you will have a stimulus that is an inverted version of another one. Perceptually these things are invariant. You would expect the cells to exhibit this kind of invariance.

Tanifuji: If the upper part and lower part are the same, then rotating upper and lower is entirely the same, so there is no difference. But if the upper part is smaller than the larger part, if you rotate the upper or lower part there is a drop of activation.

Esteky: From the data reported by Tanaka et al (1991) using simplified images like the ones used here, the orientation tuning bandwidths of IT cells are around 100 degrees or so. This means that IT cells usually won't respond to the inversions (i.e. 180 degree rotation) of those simplified images.

Diamond: In a sense that kind of variance, in contrast to invariance, is a kind of classification. When there is a small ball on top of a square it might look like head and shoulders, but when it is rotated it is not head and shoulders any more.

Albright: It's an upside down head and shoulders. If you saw a person upside down you would still recognize them as a person.

Logothetis: It has been disputed. What you are referring to is Charlie Bruce's study. He ascribed this to some kind of lateralization in the monkey brain. He showed that they don't care so much if it is upside down. However, if you make all conditions similar to the conditions used in human experimentation, there is a face inversion effect in monkeys.

Derdikman: As a sort of control, what happens if you try to apply the object simplification method to a face area? You wouldn't expect this procedure to work. In the face area we would expect the features to be very important, and if they are lost the response is likely to cease.

Tanifuji: In the area we recorded, some neurons are activated by faces. Some of these neurons responded to a visual feature of faces. For example, some neurons are activated by a monkey face, but also by a bottom view of an apple and a concentric circle. The concentric circle, where the central part and surrounding part have different brightness, is a common feature among these stimuli.

Now, there are the other group of cells that seem to represent face configuration. For these cells, it is usually very difficult to do simplification. If we remove any part of the face stimulus, the response drops down to some extent. We can also do optical imaging with faces too. In our previous study (Wang et al 1998), we found that a face activates multiple spots. Many of these spots appeared when a face was presented in a particular view angle. Perhaps this is because a visual feature involved in one view disappeared in the other view. In addition to such spots, we found that faces with different view angles activated spots that are partially overlapping. These spots are systematically arranged along the cortical surface from left profile, front face, to right profile. There is some evidence from extracellular recordings that this is the region where face configuration is represented.

Thus, to answer your questions, there are spots specific to faces, and it is difficult to apply stimulus simplification, but this is not because some features are face specific, but that the spots represent facial configuration that will be broken if some visual features are removed.

Haggard: Many of your stimuli are bodies. You are emphasizing the idea of a syntactic spatial structure, such as 'on top of'. Of course, the parts of the body move relative to one another an enormous amount. It seems odd that bodies should be encoded in terms of spatial relationships of their parts.

Tanifuji: We are looking at geometrical structures, and they do not necessarily need to be related to bodies. I would also like to mention that the 'on top of' is one particular configuration that is easy to address. There could be many others, and various configurations of bodies are represented by them.

References

Tanaka K, Saito H, Fukada Y, Moriya M 1991 Coding visual images of objects in the inferotemporal cortex of the macaque monkey. J Neurophysiol 66:170–189

Fujita I, Tanaka K, Ito M, Cheng K 1992 Columns for visual features of objects in monkey inferotemporal cortex. Nature 360:343–346

Wang G, Tanifuji M, Tanaka K 1998 Functional architecture in monkey inferotemporal cortex revealed by in vivo optical imaging. Neurosci Res 32:33–46

General discussion IV

Romo: Most of our knowledge about how cortical circuits function comes from single unit studies in perceptual or motor tasks. We always wonder whether the columnar organization described for the sensory cortex applies also to more central areas. Giaccomo Rizzolatti, have you ever observed mirror columns?

Rizzolatti: Before discussing mirror neurons, I would like to raise a more general issue on the columnar organization of the cortical visual areas. My view is that what we call columns in the middle temporal visual area (MT) and what we call columns in the inferior temporal cortex (IT) are entities that are conceptually very different. In the case of MT, the cortical columns code specific stimulus features, such as visual stimulus direction. In the case of IT, the columns code various, complex visual features that, taken together, represent an even more complex, sometimes meaningful, visual representation (e.g. a face). In one case what is coded are different aspects of the same parameter; in the other visual features that may largely differ from one another in the different neurons of the same column. I don't expect, therefore, that the response coherence within columns of different areas should be the same. Intracolumnar coherence should be very high in primary visual cortex and MT (as it is), and it should be low in IT.

Albright: Empirically, this seems to be the case. In the earlier areas, things are arrayed parametrically across the cortex. You could make the argument that in IT we just don't know what the right parameter is. If you manipulated the right frame you might find this parametric change across the cortex and it wouldn't look like these isolated categorical representations. It is hard to imagine that there is a discrete boundary between the chunk of cortex that is representing a face and chunk that is representing a fire hydrant.

Rizzolatti: Do you agree, therefore, that organization logic should be different?

Albright: I agree with the data!

Logothetis: In one case it was quite easy to find one parameter in all this chaos, and express things along this dimension and get some kind of cortical architectural organization. In the other case we had absolutely no clue as to what the appropriate dimensions are. It is characterized by a certain arbitrariness which is a nuisance. One thing we could speculate is that we have all these fancy tricks such as multidimensional scaling, and one could try to do a general classification based on some kind of metric derived from human similarity or dissimilarity measurements. This could be an alternative to chopping heads off dollies: we don't see heads moving around apart from on bodies.

Rizzolatti: No, we don't, but your focal attention is sometimes on the face, sometimes on the body.

Logothetis: Remember, these are anaesthetized monkeys.

Rizzolatti: Before being anaesthetized the monkey was not anaesthetized!

Romo: Do you see mirror columns in the motor cortex?

Rizzolatti: We never studied this issue properly.

Treves: I would like to try to reformulate your viewpoint, to make sure I understand it correctly. The spots in IT are clearly there: no one disputes the results. But they are a kind of a distraction. They are something that the experimenter is looking for; an organizing principle for interpreting the results. But it is a component of the results that is not so interesting for understanding information processing by that cortical region. The component, instead, that gives to that spot its information processing capability is the variety of responses of the neurons that respond within that spot. The mass action that we see with the optical signal is precisely what our brain has to subtract out—it is the component that perhaps is determined just by the general connectivity between different bits of cortex. It is important to understand the true nature of these putative columns. This can shift our attention from within-column commonalities to within-column variance.

Barash: IT is special. Not only is it difficult to sort out in terms of stimulus dimensions, but also it is probably the only major association cortex that doesn't have connections with the cerebellum, as Mitch Glickstein showed. This is my speculation: if you think of the cerebellum doing calibrations, it can take anything which is intrinsically quantitative and calibrate it and change things precisely. Why does IT not have connections to the cerebellum? Perhaps this is because it works in a different manner. Intuitively we tend to take MT and use what is correct for MT for IT. It is really different. It has to have some kind of fuzzy logic; a different computational mechanism is used here.

Logothetis: I am afraid that you are right, and this makes working in IT very difficult. One researcher shows the face and removes the eyes, and the response is gone. Then he shows the eyes only and the response is still gone. We have seen identical things with paper clips: you can do the same things we do with faces with paper clips. It was an absurd thing to do but we did it anyway. The monkeys became super experts. They could recognize paper clips from different views, which was an impossible task for any untrained human subject. After a year of doing this, in the anterior mediotemporal sulcus (AMTS) there were cells that were fine tuned, but most of them had a striking non-linear property. We would remove one piece, or one angle and the cell would stop. Then we presented it by itself, and there were still zero spikes. Some of the neurons are tuned to characteristics, and this shape information is just another characteristic. In our recordings we never saw any systematic gathering of anything we could visually classify as columns. Most of the recordings were coming from the posterior bank of the AMTS, but this was an area

8–9 mm multiplied by 3 mm. This makes quite an area. Within this area there was a striking variability from cell to cell.

Albright: The point was made that this cortex is very different, and I agree from my personal experience. Nonetheless, we still don't know what the relevant dimensions are, so it is sort of like a fishing expedition. When we go into MT, we know exactly what the cells respond to, so we can characterize them spatially across the cortex. In IT we just don't know what to manipulate. The things we have been able to manipulate have been based on our intuitions about object recognition. It may come to pass that if we have the right dimension and manipulate it, we will find there is some sort of systematic representation across the cortex. The whole thing will be a lot more appealing to study.

Barash: Do you have any idea why this is the only major association cortex not connected to the cerebellum?

Albright: It is involved in recognition rather than action. But this is just hand waving.

Diamond: Tom Albright described it as a fishing expedition. But isn't this just the sort of problem that Uri Hasson's reverse correlation in previewing might help resolve?

Albright: It would be hard to interpret the data. You could do reverse correlation, and the standard way of doing that with single-cell recording is to compute the spike-triggered average stimulus, but these responses are so non-linear you aren't going to get anything meaningful out of it.

Diamond: Just to find the dimension. It is not a coding problem; it is finding the sort of stimulus that works best.

Logothetis: Tom Albright has stressed that we don't know the dimensions of the stimulus. With our paper clip stimuli, they did not have too many dimensions except of configuration of information. They were deliberately selected with the same thickness, length and space. Everything was the same so we could stress only the configuration. It was like taking the same wire and bending it in different ways. In theory you could expect that if things that are similar are in one particular location there would be more of a gathering of neurons. The counter argument is that we selected an incredibly unnatural thing. If you added all the wires we tested together you would get a cloud. There is no prototype. We hardly ever learn things that don't have some kind of prototypical form.

Esteky: Going back to the main theme of this meeting, decision making and action, I think it would be useful to make a distinction between the type of percept that leads to action, and those which normally don't. For example, for a monkey, a human or a monkey face is a percept with social value. Such percepts are associated with competition or reward but a giraffe face is not related with such social or biological importance. We know that IT cells that respond to primate faces also respond to non-primate animal faces. So a single IT cell signals the presence of both types of visual stimuli. The question is how these different types of information are sent to different downstream brain sites from an identical cell assembly. A solution is that behaviourally salient neural information is channelled to appropriate brain sites for decision making and action depending on the time course of neural activity using time gating mechanisms. Such time gating requires a temporal latency code discriminating different types of objects based on their saliency. Our finding that many IT cells respond to primate and non-primate animal faces with comparable magnitudes, but more quickly to human faces than to non-primate animal faces may be taken as evidence for such a time-dependent gating mechanism. Theoretical models for such time dependent saliency mapping have been proposed (e.g. Van Rullen 2003).

Schall: Daniel Wolpert, the reaching reactions and adjustments you described are at a subconscious level. We are not aware of them. If we try to do these at a conscious level, does our performance suffer?

Wolpert: That's an interesting question. We are considering doing the cognitive version of that task where we have some discrepancy and have to estimate it. My guess is that it is not a unitary cognitive measure. From intuition, it seems that when we try to think about our feet as we are going downstairs, that is when we trip up. Thinking often doesn't help in motor control. There is a lot of evidence from game theory that people don't behave optimally in decision making. Do those game theories have the statistics of the world which we operate in, or are they arbitrary statistics which we don't have the right priors for?

Haggard: We have some quite relevant data (Johnson et al 2002). We studied the double step paradigm, with manual reaching. The subject reaches for a target and the target suddenly jumps to one side. We know that people can make rapid corrections to the new target location. We asked the subject immediately after they had made each trial to then try to reproduce the movement trajectory that they thought they had just made. It is like an enacted report of their conscious experience of their own movement. We found that whereas the motor corrections in the original movement were quick and rapidly directed towards the moving target, when people repeated the movement they just made they showed that they thought they had made the correction later than they actually did. The interesting contrast was with an anti-reaching task, where the target jumps to the right and you have to move to the left. There, people thought they were much better than they actually were. This was consistent with the idea that there are two mechanisms, one of which is partly below the level of conscious awareness and is about goal-oriented behaviour, and one of which is about our knowledge of own intentions. The latter is rather over-optimistic in the sense that it represents what we intended to do, or what we were supposed to do, rather than what we actually did.

Scott: Statistics of experience is an important issue. Mathew Diamond has been talking about this with respect to the visual system. On the motor side we need statistics about priors and all that related information. However, one of the significant things about the motor system is that we learn things and then are able to generalize and extrapolate to new situations. How does this fit in with this natural bias of statistics behaviour?

Wolpert: There are many ways you can generalize in a task. People have examined generalization of visual and motor learning. I suspect that you want to generalize in a way that the system naturally varies. We have looked at coding visuomotor learning: how we point in space. How does it generalize? It seems to generalize in head-centred coordinates. This makes sense because miscalibrations are going to come up in those coordinates. My guess is that the brain has evolved to generalize and coordinate systems that make sense of external tasks and inaccuracies in our senses. These patterns of generalization may be hard-coded to a large extent. Statistics of the world determine that: statistics over the lifespan and over evolution.

Treves: Did you say some of these are hard-wired?

Wolpert: Yes. If we learn a dynamic force field we generalize in intrinsic coordinates. We adapt to prisms easily but find it hard to adapt to a rotational space away from the body, such as rotating a computer mouse 90 degrees and trying to use it. It may be easy to adapt to prisms because our eyes rotate about the axis that prisms rotate vision about. Therefore prismatic distortions can be considered a natural miscalibration.

Logothetis: You all know the trick with the three men sitting in a row with two black hats and one white hat. Each can only see the one in front. One person has to guess their own colour and if they guess wrongly they die. Who should answer? It is the middle one, but he can only answer correctly if he waits. If the person behind who can see both others doesn't say anything, then the middle man knows that his hat is different from the person's in front; that the man at the back sees one black and one white. Sometimes the fact that something is not happening is important information. What sort of decision-making scheme takes into account time?

Wolpert: I think it's vitally important. In motor control people are very sensitive to events that don't happen. For example, if you try to lift an object up and it doesn't lift up when you thought it would.

Logothetis: I am thinking of something slightly projecting into the future. You have to integrate over time, beyond the point at which you are asked to make a decision.

Derdikman: Your solution to the hat problem only works if no one makes mistakes and everyone is super-intelligent.

Logothetis: Bayesian estimates also work only if you are very intelligent. Brains are supposedly intelligent no matter how dumb they are. Bayesian estimates are very clever estimates.

Barash: With regard to cognitive illusions, my impression is that there are many types of decisions. Bayesian inference may be used in some occasions and not in others. Your argument is strong, but it may be stronger if you don't try to argue for optimality.

Wolpert: I don't know enough about cognition to make claims that what we do cognitively is Bayesian.

Barash: If you say that motor control is optimal in all conditions you are drawing away from cases in which it is clearly helpful.

Wolpert: We have limited resources, so we can't be optimal at everything. Let's say you are bad at something: it may be because you don't do it often and don't assign many resources to it. If you have limited resources and do thousands of tasks, if you choose to assign resources equally to all tasks you will not be perfect at all of them. If you choose to assign more resources to tasks that are most important then you will have some which you can't optimise because you don't have enough neurons associated with them. It seems that we can't do everything.

References

Johnson H, Van Beers RJ, Haggard P 2002 Action and awareness in pointing tasks. Exp Brain Res 146:451–459

Van Rullen R 2003 Visual saliency and spike timing in the ventral visual pathway. J Physiol 97:365–377

Psychophysical investigations into cortical encoding of vibrotactile stimuli

Justin A. Harris

School of Psychology, University of Sydney, NSW 2006, Australia

Abstract. Neurons in primary somatosensory cortex (S1) respond to vibrotactile stimuli by firing in phase with each cycle of the vibration. We have investigated how neural activity in S1 might contribute to people's perception of vibration frequency. The contribution of S1 was confirmed using transcranial magnetic stimulation (TMS): accuracy in comparing the frequency of two sequential vibrations was reduced by a single TMS pulse delivered to S1 in the interval between the two vibrations. More recent experiments have revealed that participants use the velocity (or energy) of the stimulus when judging its frequency. This is consistent with a contribution from S1: electrophysiological recording in S1 cortex of rats shows that neurons in S1 do not explicitly code vibration frequency, but instead code the product of frequency and amplitude (proportional to the mean velocity or energy of the vibration). Further, frequency discrimination is reduced by the addition of even very small amounts of noise to the temporal structure of the vibrations (making them irregular). However, noise has no effect if the two vibrations are presented on opposite fingertips (i.e. beyond the range of receptive field sizes of neurons in S1), or if there is no difference in their velocity. Therefore, when judging vibration frequency, humans utilize information about stimulus velocity as coded by neurons in S1, but this coding is dependent on the temporally regular input of the vibration.

2005 Percept, decision, action: bridging the gaps. Wiley, Chichester (Novartis Foundation Symposium 270) p 238–250

Measuring perceptual judgments about stimuli while manipulating their physical characteristics can uncover which aspects of the sensory signal are relevant to perception and reveal the algorithms involved in sensory processing. The present work addresses the nature of stimulus coding in the somatosensory system, in particular the processing of vibrotactile stimuli. In primates, rapidly adapting neurons in subcortical stations and the primary somatosensory cortex (S1) respond to low frequency vibrations (below 50 Hz) by firing in phase with each cycle of the vibration (Mountcastle et al 1990a, 1967, 1969). Does this activity correspond to a distinct percept that can be used in making decisions about the stimulus? To answer this question, Ranulfo Romo and colleagues have measured neuronal activity from the cortex of monkeys engaged in a task to compare the frequency of two sequential vibrations (see accompanying chapter by Romo in this volume). Their observations revealed that the second somatosensory cortex (S2) and areas of frontal cortex (prefrontal cortex and medial premotor cortex) each play an important role in the task, but suggested that activity in S1 may *not* be essential other than by furnishing a reliable signal to S2 (Romo & Salinas 2003, 2001). However, our recent psychophysical experiments with humans indicate that S1 *does* contribute to both mnemonic and comparison components of a frequency discrimination.

The contribution of S1 to a tactile perceptual task can be investigated by manipulating the topographic locations of stimuli (Harris et al 2001a, 1999, 2001b, Diamond et al 1999). Given the somatotopic organisation of S1, where neurons have small receptive fields confined to digits on the contralateral hand (Shoham & Grinvald 2001, Merzenich et al 1978, Iwamura et al 1993), S1 neurons could be expected to support the comparison between two successive vibrations only when both vibrations are presented to the same fingertip. Comparisons between vibrations presented to opposite fingertips would be expected to depend on neurons with bilateral receptive fields, such as are present in S2 (Jiang et al 1997, Gelnar et al 1998, Disbrow et al 2001, Maldjian et al 1999a,b). To investigate the role of S1 in frequency comparison, we asked human subjects to compare the frequency of two vibrations presented to either the same or to different fingertips (Harris et al 2001c). Subjects became less accurate when we increased the somatotopic distance between the comparison vibrations (presenting the two vibrations on different fingertips). This drop in accuracy could result from the exclusion of S1 from the task because S1 neurons have somatotopically restricted receptive fields.

In a subsequent study, we used transcranial magnetic stimulation (TMS) to more directly investigate the role of S1 in the comparison of vibration frequency (Harris et al 2002). Again, participants were required to compare the frequencies of two vibrations separated by a fixed retention interval. Their performance was significantly disrupted when we delivered a single TMS pulse to the contralateral S1 early (300 or 600 ms) in the retention interval. TMS did not affect tactile working memory if delivered to contralateral S1 late in the retention interval (at 900 or 1200 ms), nor did TMS affect performance if delivered to the ipsilateral S1 at any time point (see Fig. 1). Primary sensory cortex thus seems to act not only as a centre for on-line sensory processing, but additionally as a transient storage site for information that contributes to working memory.

More recent experiments extend our previous work by investigating how S1 might contribute to the perception of vibration frequency in humans. One possibility we have examined is that precise information about frequency is represented in the temporal structure of the phase-locked activity of S1 neurons. For example, the frequency may be coded by the regular time intervals separating spikes evoked in S1 (Mountcastle et al 1990b). If subjects use this information, then the addition

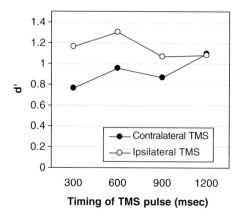

Timing of TMS pulse (msec)

FIG. 1. The effects of transcranial magnetic stimulation (TMS) on peoples' sensitivity at discriminating vibration frequency. Subjects were required to compare the frequency of two vibrations presented sequentially to the same fingertip (left or right index finger). During the 1500 ms retention interval between the vibrations, a single pulse of TMS was delivered over either the right or left primary somatosensory cortex at either 300, 600, 900 or 1200 ms into the interval. The subjects became less accurate when the TMS pulse was contralateral to the vibrations (i.e. the vibrations were on the left index finger and the TMS was presented to the right S1, or vice versa) and delivered early in the retention interval. TMS did not affect performance when delivered late in the interval or when delivered to the S1 cortex ipsilateral to the vibrations. Discrimination accuracy is presented here as d' scores.

of noise to the temporal structure of the vibration (i.e. making them irregular) will create corresponding noise in the activity of S1 neurons, and so interfere with the perception of frequency. Romo et al (1998) tested this very hypothesis but found that noise had no effect, leading them to conclude that their monkeys do not use spike timing information in S1 to code for frequency. But given other evidence that S1 does contribute to the frequency comparison in our human subjects (unlike the monkeys studied by Romo and colleagues), it remains possible that our subjects use the regular timing of spikes in S1 to code for frequency.

To examine the influence of noise on frequency discrimination, we asked participants to compare the frequency of two vibrations while we parametrically varied the level of noise added to the vibrations (Harris et al 2005). One experiment assessed performance when the available interval length information was systematically varied by the inclusion of 2, 4 and 8% noise. A second experiment extended this range of values to 9, 12, 15 and 18% noise. These values of noise are defined as the SD of the distribution of interval lengths as a percentage of the mean of the distribution. For example, a 20 Hz vibration with 8% noise has a mean interval length of 50 ms and a SD of 4 ms.

The addition of noise to the vibration reduced the subjects' sensitivity in discriminating the frequency of two vibrations delivered to the right index fingertip.

The effect is shown in Figure 2, in which accuracy scores for the frequency comparison are converted to d′ values because these conform to a more linear scale of discrimination sensitivity. A striking feature of these results is that increasing amounts of noise had little additional effect on performance. Adding as little as 2% noise reduced the d′ score by about a third, but adding larger amounts of noise (up to 18%) had equivalent effects on performance. The results suggest that a 'floor effect' prevented performance from dropping below a particular level. This has led us to speculate that performance in comparing vibrations results from a combination of two independent processes, only one of which is degraded by noise; the other process can support a relatively constant level of performance despite increasing amounts of noise. The capacity of this process to withstand the effects of noise presumably indicates that its coding of frequency is sufficiently noisy as to dilute the impact of noise added to the vibration.

Another experiment examined whether the effect of noise on frequency judgements was specifically related to the contribution made by S1 to the task (Harris et al 2005). This was achieved by manipulating the location of the vibrations being compared (either presenting both to the same index finger, or presenting one to the left index finger and the other to the right index finger) while manipulating the temporal structure of the vibrations—they were either both regular or both contained 20% noise. The two manipulations were combined in a 2 × 2 factorial design, creating four conditions and thus allowing us to determine whether the effect of noise differed for same-finger versus opposite-finger comparisons. The experiment confirmed our previous finding that when the two vibrations were presented to the

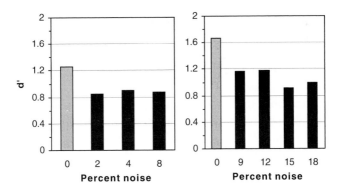

FIG. 2. The impact of noise on subjects' accuracy (shown here as d′) in comparing vibration frequencies. The amount of noise added to the vibrations was varied, and is expressed as the SD of the Gaussian as a percentage of the mean inter-deflection interval. For example, a 20 Hz vibration with 4% noise has a mean inter-deflection interval of 50 ms and an SD of 2 ms. Adding as little as 2% noise reduced discrimination performance, but adding larger amounts of noise had little additional effect.

same finger, accuracy was reduced by about one third when the vibrations were noisy ($d' = 0.80$) compared to when they were regular ($d' = 1.18$). Importantly, when the vibrations were presented on different fingers, accuracy was lower overall and did not change depending on whether the vibrations were regular or noisy ($d' = 0.80$ and 0.82, respectively). Therefore, the effect of noise was confined to comparisons between vibrations presented on the same finger, and thus may be specific to the contribution from S1 neurons. In contrast, noise did not affect comparisons between vibrations on different fingers that presumably depend on neurons with bilateral receptive fields, such as S2 or frontal cortex.

The second possibility regarding how vibration frequency is encoded has its roots in recent electrophysiological experiments conducted in the laboratory of Mathew Diamond. Arabzadeh et al (2003) recorded neuronal activity in S1 cortex of rats while simultaneously delivering vibrations to the rats' whiskers. They found that the firing rate of S1 neurons changed with variations in the frequency or amplitude of the vibration, but that the S1 neuronal activity did not explicitly encode either the frequency or amplitude. Rather, the S1 activity explicitly encodes the product of frequency and amplitude, which corresponds to the mean velocity of the vibration, or properties that are proportional to velocity, such as energy (proportional to velocity^2). Therefore, information about velocity or energy, as coded in S1, may contribute to the frequency judgement made by our human subjects.

To test this idea, we asked participants to compare the frequency of two vibrations while we manipulated the amplitude of the vibrations to dissociate their frequency from their velocity or energy (Harris et al 2005). For example, the amplitude of an 18 Hz vibration was increased by 10%, whereas that of a 22 Hz vibration was decreased by 10%, so that both had the same velocity as the 20 Hz vibration with which they were compared. At the same time, we manipulated the presence versus absence of noise (10%) in the temporal structure of the vibration to determine whether there was any interaction between the two factors. As in previous experiments, the presence of noise reduced accuracy ($d' = 0.84$) below that obtained when both vibrations were regular and had equal amplitude ($d' = 1.07$). Further, accuracy was also decreased when the two vibrations had different amplitudes so as to match their velocity, and for these vibrations it did not matter whether they contained noise or not ($d' = 0.78$ and 0.75, respectively). Therefore, accuracy in comparing vibrations was reduced either by removing the information present in the vibrations' velocity or by degrading the information present in the individual interval lengths by the addition of temporal noise. The fact that combining these manipulations did not reduce accuracy more than either manipulation on its own suggests that they act on the same mechanism.

A picture emerges from the results described thus far. According to this picture, the discrimination of vibration frequency by humans depends in part on the coding of the stimulus velocity or energy, as is exhibited by neurons in S1. Further, people's ability to discriminate vibration frequency is very sensitive to the presence of noise in the temporal structure of the vibration. Since noise did not affect discrimination between vibrations that differed in frequency but not velocity, and also had no effect on vibrations presented to different fingers, we can conclude that the presence of noise specifically reduces the coding of vibration velocity in S1. The natural interpretation of these results is that noise in the temporal structure of the vibration interferes with frequency discrimination by making the coding of vibration velocity in S1 noisy. However, further experimental data rule out this interpretation. Specifically, we have found that discrimination accuracy is reduced by noise in the temporal structure of the vibration but is not affected by an equivalent amount of noise in the amplitude of the vibration (Harris et al 2005). This is important because both types of noise would create equal amounts of noise in the velocity or energy of the vibration. Therefore, the effect of adding noise to the temporal structure of the vibration does not reduce frequency discrimination simply by inducing an equivalent amount of noise in the vibration velocity.

Other evidence from the experiments described above confirms that the effect of noise in the temporal structure of the vibration goes beyond the stimulus noise itself. We used signal detection theory (Green & Swets 1966) to estimate the amount of noise present in the sensory processing of the vibration and to determine the impact of noise present in the stimulus. Obviously this 'internal noise' could arise anywhere in the sensory processing pathway, or the subsequent decision and response processes. But in order to compare the amount of this internal noise with the noise that we added to the vibration externally, we referred the internal noise to an equivalent noise in the deflection timing of the vibration. We used the subjects' accuracy (as d' scores) on trials with regular (non-noisy) vibrations to estimate the amount of internal noise. In the same way, we used the subjects' d' scores on trials with 2% noise added to the vibration to estimate the total amount of noise present on these trials. If the internal and external sources of noise are independent, they should combine additively (i.e. total noise should equal internal noise plus noise added to the vibration). However, in our experiment, the total noise estimated from the subjects' performance with 2% noise added to the vibration was much greater than that expected from the simple sum of the internal and external noise. Clearly, then, our assumption that internal and external noise are independent is wrong—the presence of external noise in the stimulus must increase the level of internal noise in the sensory representation. A reasonable explanation for this is that the fidelity of the neuronal representation in S1 is enhanced when the vibration is regular, perhaps through a tuned adaptation whose mechanism depends on the regularity of the vibration. For example, the S1 neurons that respond to the vibration might form part of a resonant circuit such that temporally regular input might entrain the circuit and reduce the variability in the activity of those neurons.

Our conclusion that S1 contributes to the frequency discrimination process contrasts with the conclusions reached by Ranulfo Romo and colleagues that S1 neurons are not actively involved in the process of remembering and comparing vibration frequency (Salinas et al 2000, Romo & Salinas 2003). Indeed, our finding that noise affects frequency comparison is also at odds with their previous studies in which monkeys compared vibrations in a protocol very similar to that employed here. In those studies, the addition of noise to the vibrations had no effect on the monkeys' psychophysical performance (Romo et al 1998). However, there are a number of differences between their studies and ours. One very pertinent difference is that, in the studies with monkeys, the two vibrations being compared always had different amplitudes—the amplitudes were adjusted so that the two vibrations had the same 'subjective intensity' (approximately equal to amplitude × frequency). Clearly this corresponds to the adjustment we have made to match the velocity or energy of the vibrations. Importantly, this manipulation reduced our participants' accuracy in comparing the vibrations, and eliminated the effect of noise. Specifically, this resolves the discrepancy between our studies and those of Romo and colleagues: our subjects use differential information about the mean velocity or energy of the vibrations to judge their frequency, whereas that information was not available in the task presented to the monkeys.

In summary, we have shown that people's ability to discriminate vibration frequency involves coding of the vibrations by neurons in S1. We have also found that the coding by S1 neurons is highly sensitive to noise in the temporal structure of the vibration. Finally, we find that this noise-sensitive process specifically codes for the velocity or energy of the vibration rather than explicitly representing the frequency. However, the contribution made by S1 neurons constitutes only part of the representation of frequency—people can still discriminate vibration frequency, albeit with lower accuracy, when information about velocity or energy is removed, when large amounts of noise are added to the vibration, and when S1 neurons are eliminated from the vibration comparison by presenting the vibrations to fingers on different hands. Electrophysiological studies in monkeys have mapped out the cortical circuitry beyond S1 that underlies this discrimination ability—it includes neurons in S2, medial prefrontal cortex and premotor cortex (Romo & Salinas 2003).

Acknowledgements

This work was supported by grants from the Australian Research Council (DP0343552) and the James S. McDonell Foundation.

References

Arabzadeh E, Petersen RS, Diamond ME 2003 Encoding of whisker vibration by rat barrel cortex neurons: implications for texture discrimination. J Neurosci 23:9146–9154

Diamond ME, Petersen RS, Harris JA 1999 Learning through maps: functional significance of topographic organization in primary sensory cortex. J Neurobiol 41:64–68

Disbrow E, Roberts T, Poeppel D, Krubitzer L 2001 Evidence for interhemispheric processing of inputs from the hands in human S2 and PV. J Neurophysiol 85:2236–2244

Gelnar PA, Krauss BR, Szeverenyi NM, Apkarian AV 1998 Fingertip representation in the human somatosensory cortex: an fMRI study. Neuroimage 7:261–283

Green DM, Swets JA 1966 Signal detection theory and psychophysics. Wiley, New York

Harris JA, Petersen RS, Diamond ME 1999 Distribution of tactile learning and its neural basis. Proc Natl Acad Sci USA 96:7587–7591

Harris JA, Harris IM, Diamond ME 2001a The topography of tactile learning in humans. J Neurosci 21:1056–1061

Harris JA, Petersen RS, Diamond ME 2001b The cortical distribution of sensory memories. Neuron 30:315–318

Harris JA, Harris IM, Diamond ME 2001c The topography of tactile working memory. J Neurosci 21:8262–8269

Harris JA, Miniussi C, Harris IM, Diamond ME 2002 Transient storage of a tactile memory trace in primary somatosensory cortex. J Neurosci 22:8720–8725

Harris JA, Fairhall AL, Diamond ME 2005 Frequency judgements of periodic and noisy vibrotactile stimuli: implications for cortical encoding. Submitted

Iwamura Y, Tanaka M, Sakamoto M, Hikosaka O 1993 Rostrocaudal gradients in the neuronal receptive field complexity in the finger region of the alert monkey's postcentral gyrus. Exp Brain Res 92:360–368

Jiang W, Tremblay F, Chapman CE 1997 Neuronal encoding of texture changes in the primary and the secondary somatosensory cortical areas of monkeys during passive texture discrimination. J Neurophysiol 77: 1656–1662

Maldjian JA, Gottschalk A, Patel RS, Detre JA, Alsop DC 1999a The sensory somatotopic map of the human hand demonstrated at 4T. Neuroimage 10:55–62

Maldjian JA, Gottschalk A, Patel RS et al 1999b Mapping of secondary somatosensory cortex activation induced by vibrational stimulation: an fMRI study. Brain Res 824:291–295

Merzenich MM, Kaas JH, Sur M, Lin CS 1978 Double representation of the body surface within cytoarchitectonic areas 3b and 1 in "SI" in the owl monkey (Aotus trivirgatus). J Comp Neurol 181:41–73

Mountcastle VB, Steinmetz MA, Romo R 1990a Frequency discrimination in the sense of flutter: psychophysical measurements correlated with postcentral events in behaving monkeys. J Neurosci 10:3032–3044

Mountcastle VB, Steinmetz MA, Romo R 1990b Cortical neuronal periodicities and frequency discrimination in the sense of flutter. Cold Spring Harb Symp Quant Biol 55:861–872

Mountcastle VB, Talbot WH, Darian-Smith I, Kornhuber HH 1967 Neural basis of the sense of flutter-vibration. Science 155:597–600

Mountcastle VB, Talbot WH, Sakata H, Hyvarinen J 1969 Cortical neuronal mechanisms in flutter-vibration studied in unanesthetized monkeys. Neuronal periodicity and frequency discrimination. J Neurophysiol 32:452–484

Romo R, Salinas E 2001 Touch and go: decision-making mechanisms in somatosensation. Annu Rev Neurosci 24:107–137

Romo R, Salinas E 2003 Flutter discrimination: neural codes, perception, memory and decision making. Nat Rev Neurosci 4:203–218

Romo R, Hernández A, Zainos A, Salinas E 1998 Somatosensory discrimination based on cortical microstimulation. Nature 392:387–390

Salinas E, Hernández A, Zainos A, Romo R 2000 Periodicity and firing rate as candidate neural codes for the frequency of vibrotactile stimuli. J Neurosci 20:5503–5515

Shoham D, Grinvald A 2001 The cortical representation of the hand in macaque and humans S-I: high resolution optical imaging. J Neurosci 21:6820–6835

DISCUSSION

Schall: Did you test different digits across the hand? Is there a homologous finger effect?

Harris: No. At longer retention intervals this happens, though. At shorter retention intervals it has to be the same finger or performance is lower. At longer intervals we start to see a symmetrical pattern, where the corresponding fingers do better and non-matching fingers on the same hand do worse.

Logothetis: You showed that it is not the frequency but the velocity that is encoded. Then you say that the noise is independent; that the noise in velocity doesn't affect behaviour.

Harris: The effect we see with this noise in the temporal structures doesn't seem to be happening as a result of making the coding of velocity noisy.

Logothetis: Could it be that in reality what matters is frequency, and amplitude is only important in activating sufficient numbers of receptors? As you increase the amplitude you get less sensitive receptors also signalling.

Harris: Are you suggesting that increasing amplitude would reduce frequency discrimination?

Logothetis: No, it would increase it to some extent, then there could be a cut-off. This would make a system that is sensitive to frequency, although it gives you the impression that it is sensitive to velocity. It is independent of the noise in velocity. After you increase the amplitude by a certain amount it will activate as many receptors as possible: from that point on what is really encoded is the frequency.

Harris: So you are suggesting that the amplitude we made to match velocity was somehow bridging that critical cut-off; that when we reduce amplitude by a higher frequency vibration we are falling below the level that you say is critical.

Logothetis: This is just my speculation. Usually, if you encode something you are sensitive to noise in the particular thing you are encoding. It is surprising that the noise in the amplitude didn't have an effect.

Harris: Yes, I agree: I was expecting it to do that.

Logothetis: You can define the velocity either as amplitude divided by the frequency or the temporal frequency divided by the spatial frequency. In your case the spatial frequency isn't going to change.

Harris: The condition in that experiment in which we had temporal noise and amplitude noise, was that those two sorts of noise were anti-correlated. Wherever the interval was shorter and our instantaneous frequency was higher, the amplitude was reduced by the same amount. The instantaneous velocity of those vibrations was constant. The idea was that this might reinstate performance. This didn't occur. I guess there are problems about whether amplitude and frequency are being transmitted and encoded in-synch enough for this effect to hold out.

Diamond: The hypothesis that the noise in the stimulus augments the internal noise is interesting. This effect occurred for frequency noise and not for velocity noise. Do you think the internal noise is somehow specifically sensitive to disruption more by frequency than by amplitude?

Harris: I guess this is why I was entertaining the idea that in fact there might be some sort of oscillator or feed-forward phase-lock-loop process. If the stimulus is perfectly regular it could get entrained to this and therefore improve the response to these deflections. This should be very sensitive to disruption of temporal structure and less affected by variability in amplitude, as long as you are above some sort of minimal amplitude in order to get a robust activation.

Haggard: I have a question about the memory process. You showed how transcranial magnetic stimulation (TMS) over S1 would interfere with the tactile vibration discrimination task. TMS applied shortly after the first stimulus was more effective than the TMS applied for longer after the first stimulus. In other words, did the effect of TMS change over time?

Harris: It actually wasn't any different between 300 and 600 ms.

Haggard: Memory implies something that is stored over a period of time. Therefore an intervention which aimed to neutralize or abolish that representation should be relatively time insensitive. If the information is *stored* in S1, then TMS ought to produce an equal impairment at whatever time it is delivered, up until the time information is copied out of S1 to some other location.

Harris: The implication is that if S1 is holding information about the first vibration, it is doing it fairly transiently, but that information must be leaving S1, unless it gets into a form that becomes robust against TMS.

Haggard: The question then is how do you distinguish between this sort of memory effect and a masking effect? Masking would be effective just after the stimulus and become less effective over time.

Harris: I have done the backward masking experiments with touch. If you get masking effects at intervals of 50 ms, it is almost gone by 100 ms. 300–600 ms is pretty much out of the range of masking. Also, the vibration is an entire second long.

Haggard: Is there any relation between the kind of memory you are proposing may exist in S1 and the psychological concept of iconic storage?

Harris: Given the timecourse, I am inclined to think that it is some sort of sensory memory such as iconic or echoic. I am planning to do a sort of George Sperling experiment to see whether I can document sensory memory in the tactile system and its timecourse.

Haggard: That is the best behavioural signature we have for these very short-term sensory memories.

Treves: I am just going to try to put together your model with oscillators and Nikos Logothetis' comment. The real coding might be in terms of some resonant frequencies of oscillators in S1, and perhaps earlier than S1. Then as you recruit more receptors or oscillators you might see a kind of apparent coding in terms of velocity, which still is not sensitive to noise in the amplitude because the underlying coding is the resonant frequencies of the oscillators.

Harris: Are you saying that the oscillation process, if it exists, is what is giving rise to the apparent evidence for velocity coding?

Treves: Yes, there is a distribution of oscillators at different frequencies, some of which require higher amplitudes to be activated.

Harris: If we add noise to the amplitude of the vibration, wouldn't that also disrupt this oscillator process?

Treves: No, the number of the oscillators is sensitive to the amplitude. This is not a process that is much affected by noise. Noise may disrupt entrainment of individual oscillators.

Harris: When I manipulated the amplitude of the vibrations to match their velocity, this was done by increasing the amplitude of the lower frequency vibration for half of the number of trials, and by reducing the amplitude of the higher frequency vibration on the other half of trials.

Logothetis: I was trying to suggest that you could explain these results because if you stimulate more with a larger amplitude or smaller amplitude, if there is any recruitment, you might expect a competition between things that adapt and things that are being recruited. I don't know whether this makes any sense. It is not intuitive to me how you can have a system that is encoding in one dimension and being insensitive to noise in that dimension.

Harris: It is insensitive within the realm of measurement error we have in this task.

Wolpert: You are stimulating with different amplitudes and velocities, but you are going through this low-pass system of the skin. What are you actually getting in to the sensors at the end? Are you controlling these two things independently or didn't you worry about this?

Harris: I didn't worry about it.

Wolpert: I assume the skin has lots of interesting properties.

Scott: I'm just thinking of this idea of the frequency nulls. There is this problem with noise, and you do much better when it has no noise in it. But in the real world is there ever a texture signal which has a perfect frequency? Entraining is a useful way of keeping this information, but would you ever get something that perfect?

Harris: That's a good point. The idea of some sort of adaptive process tuning to a perfectly regular vibration is a way of explaining this. The existence of such a mechanism would then have to be explained.

Diamond: You don't have to think about vibration exclusively as arising from textures. We can pick up a lot of meaningful information from vibration. It is not necessarily the same thing as texture.

Romo: Your stimulus duration is about 1 second and my comment relates to the contribution of S1 to the early component of the delay period. By the time the first stimulus ends, all the memory areas are already engaged, because the sensory input to S2 is about 5–6 ms after S1. We know that with 200 ms stimulus length you can get enough information in S1, S2 and frontal lobe. When you apply the TMS you might be affecting S1, S2 and the rest of the system. This is another reason why you might be observing some effects of the discrimination process in S1. It doesn't prove that S1 contributes to the working memory component of the task. You might be affecting the downstream representation of S1. This is one possibility. Since we started the recordings in 1987 we never observed delay responses between the two stimuli in S1 neurons. We see from time to time some responses during the delay, which might be associated with some finger motions. These responses, however, are not related to the encoding of the first stimulus frequency during the working memory period. Another possibility is that S1 of humans is different to S1 of monkeys.

Harris: I guess there is also the issue that our subjects are brought into our lab, sat down, and the task is explained to them, whereas the monkeys receive months of training.

Romo: I favour my first interpretation. By the time the first stimulus ends, all the systems are already engaged in the task.

Harris: Two responses to that point. Would you expect a time dependency effect? If the TMS was just having a downstream effect on this circuit, would you expect it to have its effect if you deliver TMS at 300 or 600 ms, but not at 900 or 1200 ms?

Romo: In that case you will affect the sensory component. So you will therefore affect the working memory component as well. There is no doubt that if you do something in S1, you might be affecting the whole system.

Harris: Why would it only have an effect if you did it to S1 in that early part of the attention interval and not later?

Romo: We have many different sets of neurons that are engaged in the working memory component. There are neurons that respond persistently, and some which respond late. These neurons might be compensating the early effect of the TMS stimulation. I have no doubt that you are producing something with your stimuli, but I wonder whether it is truly related to the working memory component of the frequency discrimination task.

Diamond: If the only evidence for S1 involvement in working memory were TMS, it would be only weakly suggestive. The argument is stronger because it is combined with topography experiments which show that people can make comparisons better for two stimuli delivered to the same fingertip than two stimuli delivered even to neighbouring fingertips on the same hand. In your interpretation that only downstream areas, such as S2 and PF, are involved in working memory, would you still predict that the relative positions of the stimuli makes a strong difference? Would you expect that the neurons in S2 or prefrontal cortex compare two successive fingertip stimuli in a different manner depending on the relative positions of the stimuli?

Romo: Discrimination thresholds in monkeys are measured when they are fully trained. Normally, our animals are trained by delivering the stimulus to only one finger. When the monkey is able to make a true comparison between two stimuli and measure what we call a normal discrimination threshold, we try another finger. The psychometric thresholds we observe are very similar to what we observe from the chronically stimulated finger. Our interpretation is that there is a mechanism downstream that doesn't care about the topography, but gives a simple reading of the stimulus making the comparison no matter where the stimulus comes from. This relates to what is happening with the prefrontal cortex.

Harris: The observation you have described is one where you are still presenting the two vibrations on the same fingertip. You should change finger.

Romo: During recording, electrodes can be located in the middle, third or fourth fingers and we have to take advantage of that. The combined neurophysiological/psychophysical experiment has to take advantage of this. We sample the response properties from day to day by recording receptive fields from different fingers. Even in microstimulation experiments, we can deliver the mechanical stimulation in the index finger and the electrical stimulation in a column of neurons that are mapping information from the fifth finger. Discrimination performance is as good as when the two mechanical stimuli are delivered to the same finger. Somewhere in the brain those two signals might be combined to make the comparison and produce the decision motor report.

Derdikman: It could be that in S1 there is no superthreshold contribution to working memory, but there are some subthreshold differences that are related to it.

Romo: That is also possible. We are dealing with spikes and we don't know what is going on with subthreshold signals.

Why is language unique to humans?

Jacques Mehler*†[1], Marina Nespor‡, Mohinish Shukla* and Marcela Peña*

*International School for Advanced Studies, Trieste, Italy, †Ecole des Hautes Etudes en Sciences Sociales, Paris, France and ‡Universita di Ferrara, Ferrara, Italy

Abstract. Cognitive neuroscience has focused on language acquisition as one of the main domains to test the respective roles of statistical vs. rule-like computation. Recent studies have uncovered that the brain of human neonates displays a typical signature in response to speech sounds even a few hours after birth. This suggests that neuroscience and linguistics converge on the view that, to a large extent, language acquisition arises due to our genetic endowment. Our research has also shown how statistical dependencies and the ability to draw structural generalizations are basic processes that interact intimately. First, we explore how the rhythmic properties of language bias word segmentation. Second, we demonstrate that natural speech categories play specific roles during language acquisition: some categories are optimally suited to compute statistical dependencies while other categories are optimally suited for the extraction of structural generalizations.

2005 Percept, decision, action: bridging the gaps. Wiley, Chichester (Novartis Foundation Symposium 270) p 251–284

Linguists and psychologists have studied language acquisition; the former have elaborated the most sophisticated formal theories to account for how this unique competence arises specifically in humans. For instance, Chomsky (1980) formulated the Principles and Parameters theory (hereafter, P&P) to account for the acquisition of language given the poverty of the linguistic data the learner receives. In fact, infants acquire the grammatical properties of their language of exposure on the basis of partial and unreliable information. Babies, like adults, are confronted with incomplete or erroneous sentences.

P&P assumes that infants are born with 'knowledge' of Universal Grammar. This endowment includes genetically specified universal principles, that is, the properties shared by all natural languages. Moreover, the endowment specifies a number of binary parameters that capture those grammatical properties that vary systematically between groups of natural languages. For instance, there are groups of lan-

---

[1] This paper was presented at the Symposium by Jacques Mehler, to whom correspondence should be addressed.

guages that put Heads to the left of Complements while other languages put Complements to the left of Heads. The P&P theory attempts to identify such structural properties that are basic to natural language distinctions. Parameters can be thought of as switches that must be set to one of two possible positions to specify the properties of the language being learned. The linguistic input determines the particular value of a parameter.[2]

P&P has many virtues. First, by exploring the way in which natural languages are sorted into groups that share coherence for syntactic properties, P&P is one of the most productive theories ever developed within the linguistic domain, (see Baker [2001] for an accessible and fascinating account of the P&P proposal). Next, P&P also addresses the problem of language acquisition without making the simplifications common to alternative theories. For example, optimists claim that imitation is the privileged mechanism responsible for the emergence of grammatical competence. The P&P perspective is appealing because it is biologically realistic assuming that infants are equipped with a species-specific mechanism to acquire natural language that can be explored with the tools available to formal linguistics and to the explorations of cognitive neuroscience.

While P&P is certainly playing an important role in the domain of language acquisition, there is a second influential position that asserts that the surface properties of stimuli can bias the learner towards postulating syntactic properties for the incoming utterances. While the P&P theory was formulated with the precision necessary to allow us to evaluate it, the general learning device proposal appears to be somewhat less precise. Criticisms of proposals according to which general learning mechanisms are sufficient to explain language acquisition have been given by many theoreticians, see Chomsky (1959), Fodor (1975), Lenneberg (1967) and Pinker (1994) among many others. We will come back to this point below.

Recently, some attempts were made to show that speech signals contain hitherto ignored information to allow general learning accounts to explain how language is acquired. Mostly, these attempts are minor modifications of association, a mechanism that humans and animals share. Within this stream of research, the brain is

---

[2]To illustrate this, consider a child who hears mostly sentences with a Verb–Object order. The child, supposedly, obtains information automatically from the linguistic input to set the relevant word order parameter. If this were so, it would constitute a great asset, since fixing this word order parameter may facilitate the acquisition of grammar and also the acquisition of the lexicon. Likewise, the child exposed to a language that can have sentences without an overt subject e.g. Italian ('piove', 'mangiano arance' etc.), or to a language whose sentences require overt mention of subjects e.g. English ('it is raining', 'they eat oranges'), supposedly gets information from the linguistic input to set the relevant pro-drop parameter.

regarded as a huge network that works in a Hebbian fashion (see Hebb 1949).[3] This may explain why many psychologists and neuroscientists have adopted a viewpoint that ignores the complexity of syntax and assumes that by focusing exclusively on speech perception and production, a functional account of how language is processed will follow. Undeniably, behavioural scientists have made great strides in the study of perception and production. Some of them believe that it is *sufficient* to study how language production and perception unfold during development to understand how syntax (or semantics) arises in the infants' mind. Of course it is easier to study speech perception in babies or animals than trying to figure out how the human brain computes syntax, semantics and pragmatics of utterances, something that animal brains cannot do. Psychologists who adopt a general learning device framework often assume that the mystery of syntax acquisition will disappear once we understand how infants just learn to extract the distributional properties of language (see Seidenberg & MacDonald 1999 among many others).

At this point we would like to point out that although there is a huge contrast between the two stances presented above there are many points of agreement as well. Investigators working in both the P&P tradition and in the general learning framework agree that some parts of grammar must be learned. Indeed, no one is born knowing Chinese, Malay or any other natural language. Each learner has to acquire the language spoken in his or her surrounds. What distinguishes the different positions is the scope and nature of learning they are willing to posit. P&P assumes an initial state characterized by 'knowledge' specific to a putative language module, i.e. Universal Grammar (UG). In contrast, general learning theoreticians assume that the initial state is characterized by learning principles that apply across the different domains in which organisms acquire knowledge. The general learning device is undeniably a powerful account to explain how organisms will use the surrounds to acquire behaviours that satisfy the organism's needs. Thus, it is an empirical issue whether just one of these theories is sufficient to explain how the human mind acquires its capacities including natural language.

Do we have accounts of how syntax can be acquired after the child has learned the lexicon of the language of exposure applying general learning mechanisms? Is there any evidence that the acquisition of syntax does not start until at least part of the lexicon is acquired? Depending on the answers we give to these questions, either the P&P model or the general learning model should be abandoned for syntax acquisition.

---

[3] Hebbian networks are a set of formal neurons synaptically connected but with connectivity values that change with functioning. If two neurons are active at the same time, the value of their connection increases. Otherwise, the value of the connection stays identical to what it was or decays.

So far, we have tried to highlight the positive aspects of both P&P and general learning mechanisms. However, problems arise with both frameworks. While the first tries to cope with language acquisition in a realistic sense the second focuses on the acquisition of, at best, toy-languages. P&P is problematic because of the many implicit assumptions that investigators make when trying to explain the acquisition of grammar. P&P was formulated with syntax acquisition in mind and researchers generally take for granted that infants, in one way or another, have already acquired the lexicon, before setting syntactic parameters. Presupposing that infants begin processing speech signals only when they start learning the lexicon justifies neglecting the study of language acquisition during the first year of life and explains why P&P investigators have mostly reported data from language production studies.

Data from animal experiments suggests that the vertebrate auditory system is optimally suited to process some of the linguistically relevant cues that speech affords. Thus, at least some properties of language could be acquired precociously from speech signals. Indeed, animals with auditory systems similar to our own tend to respond to speech patterns much like infants younger than eight months (see Kuhl 1987 and Ramus et al 2000 among many others). Apes, but also dogs, have 'lexicons' that can attain a few dozen words (Premack 1971, 1986). However, such abilities are insufficient to enable non-human animals to construct a grammar comparable to that of humans. Nonetheless, together with other pieces of evidence that we lay out below, we assume that the sensory capacity of many vertebrates licenses the processing of speech from the first year of life and, consequently, we should not neglect the aquisition that humans make during their first year. We show below that language acquisition begins with the onset of life. Indeed, several investigators, regardless of the position they defend, have found empirical evidence suggesting that the sound pattern of language are identified by very young infants and that some properties can be attested even in neonates. The sound pattern of speech contains cues that might bias language acquisition at different stages. As is becoming obvious, the viewpoints we presented above are complementary. Indeed, while rationalists and empiricists acknowledge the role of learning in language acquisition, the nature of learning conceived by each of the viewpoints is radically different. In the pages below we will try to show that it is desirable to keep in mind that only human infants use the acoustic properties of speech to acquire grammar. In order to explain how such uniqueness comes about, the theory that will eventually be preferred will be the one that fits best with biological processes.

We know that the uniqueness of syntax must be explored formally and explained with models that are biologically realistic. Indeed, we are confronting a human aptitude that will bloom under several types of impoverished learning environments. The linguistic input comes usually in the form of speech signals or, less often, in the form of hand gestures as produced by deaf humans. Whether the learner is hearing and seeing, deaf or even blind, s/he will attain a grammar that is as rich and complex as we expect it in humans without sensory filters, see Klima & Bellugi (1979) and Landau & Gleitman (1985) amongst others.

Thus, not only do we have to account for the uniqueness of the human language ability but we also have to account for how language arises despite all the described impoverished conditions. The best way to attain such an aim is to use the specifications given in P&P to explain what needs to be learned and what may be mastered throught general learning procedures.

Chomsky (1980, 1986) and others have argued that conceiving acquisition of language from a P&P perspective will bring clarity to the field. However, the mechanisms for the setting of parameters in the P&P theory were seriously underspecified so as to make it hard to judge. In fact, Mazuka (1996) argues that, in its usual formulation, P&P contains a paradox (see below). Morgan et al (1987), Cutler (1994) and Nespor et al (1996) among others, have proposed some putative solutions to some of the problems arising within the P&P proposal. However, few proposals have explored how the infant evaluates and computes the triggering signals. Some recent results suggest that two-month-olds are sensitive to the prosodic correlates of the different values of the head-complement parameter (Christophe et al 1997, 2003).

In the early 1980s, some scholars like Wanner & Gleitman (1982) already foresaw some of the difficulties in the existing theories of grammar acquisition and proposed that *phonological bootstrapping* might help the infant out of this quandary. They held that some properties of the phonological system learnt by the child may help him/her to uncover lexical and syntactic properties. Some years later, Morgan & Demuth (1996) specifically added that prosody contains signals that can act as triggers and thus help the child learn syntax. Indeed, these authors conclude, as we do above, that the study of speech signals that can act as triggers is essential if we are to understand the first steps into language.

To overcome the poverty of the stimulus argument, innate dispositions were postulated. However, as pointed out above, the proposal for language acquisition is not sufficiently specific. Indeed, if an important part of the infant's endowment comes as binary parameters, we still need to understand how these are set to values that are adequate to the surrounding language. The general assumption was that by understanding a few words or simple sentences like '*drink the juice*' or '*eat the soup*' the child would generalize that in her/his language, objects follow verbs. As Mazuka (1996) pointed out, this assumption is unwarranted. Indeed, how does the child know that soup means *soup* (Noun) rather than *eat* (Verb)? Even if the mother always says *eat* in front of different foods, the child may understand that what she means is simply *food!* If the signals were to inform the child about lexical categories or word order, one could find a way out of this paradox. Before we know if this is a valid solution, we need to ask whether such signals exist and if they do, whether the infant can process them.

The prosodic bootstrapping hypothesis arose from linguistic research that focused on the prosodic properties that are systematically associated with specific syntactic properties (e.g. Selkirk 1984, Nespor & Vogel 1986). These authors found systematic associations between these two grammatical levels, making plausible the notion that signals may cue the learner to postulate syntactic properties in an automatic, encapsulated fashion.

What is the infant learning during the first 18 months? Possibly, the answer is related to the infants' ability to perceive and categorize the cues that can act as triggers. Since these are supposed to function in an automatic and encapsulated way, supporters of the prosodic bootstrapping hypothesis are committed to the view that infants have 'learned' many aspects of their language before they begin to produce speech. These researchers have to give an account of the specific processes that occur during the first months of life. As we and others have argued, a parameter cannot be set after listening to a single utterance. Rather, properties of utterances are stored and the information is presumably used to set a parameter when it has become 'reliable'. Since some parameters can only be set after other grammatical properties have already been acquired (each of them requiring considerable information storage), we could perhaps understand the 'slow' pace of learning.

The sound of words is arbitrary as is clear from its variation attested across languages. On top of having to learn the identity of words, the child has to discover when a multi-syllabic utterance contains one or more words. Since most words are heard in connected speech, the infant has to rely on procedures to parse speech signals into its constituent morphemes. How does the infant parse speech to identify potential words? A proposal made by Saffran et al (1996) is that this can be achieved even by eight month olds, by computing the statistical properties of incoming speech signals. Thus, although we assume that UG is part of the infant's endowment and that it guides language acquisition, we also acknowledge that the statistical properties of the language spoken in the surrounds inform and guide learning. This is in contrast with the position of some theorists who argue that it is possible to explain even how grammar is acquired, exclusively on the basis of infants' sensitivity to the statistical properties of signals. How would such models stand up against real settings in which infants learn language from signals they receive? This issue was addressed by Yang (2005) who concluded that probabilities alone would not allow infants to converge to the words of the language given by the input.

The above presentation makes it clear that more data and research is needed to understand how the human biological endowment interacts with learning abilities during the first months of life. We are in a rather good position to do this because during the last few years, new and fascinating results have been secured, allowing us to start having a more coherent picture of language acquisition. We will first explore whether the brain of newborn infants is specialized for language processing or whether this specialization arises as a consequence of language acquisition.

Innate dispositions for language?

Infants experience speech in noisy environments both before and after birth. Paediatricians tend to conjecture, as do naïve observers, that the racket infants experience after birth does not interfere with the processing of speech since they learn to focus on speech during gestation. But the womb is far from being the sound-proof chamber that one might imagine; the womb is a very noisy place. Experiments with pregnant non-human vertebrates and volunteer pregnant women reveal that intra-uterine noise is as great as, if not greater than, the noise infants encounter after birth. The bowels, blood circulation, and all kinds of movements generate considerable noise (Querleu et al 1988). Thus, the womb is not the place to learn how to segregate speech from background noise. How then does the infant identify the signals that carry linguistic information? Why are music, telephone rings, animal sounds, traffic noises and other noises segregated during language acquisition?

Among the first researchers to focus on this issue we must mention Colombo & Bundy (1983) who found that young infants respond preferentially to speech streams as compared to other noises. This result, however, is difficult to appraise. There are zillions of noises out there and it is quite likely that infants might prefer some of them to the speech used by Colombo & Bundy (1983). We can always imagine that some melody is more attractive than a speech stream. Unfortunately few experiments have convincingly investigated this area. In an indirect effort, Mehler et al (1988) found that neonates and two-month olds process better, or more attentively, normal speech utterances as compared to utterances played backwards. The authors interpret their finding as showing that the infants brain preferentially processes speech, rather than non-speech stimuli. The uniqueness of this experiment resides in the numerous physical properties that these stimuli share, i.e. pitch, intensity and duration. However, in order to argue that the neonate's brain responds specifically to speech sounds rather than to the human voice (regardless of whether it is producing speech or coughs, cries or sneezes) more studies would be desirable.

Mehler et al's (1988) study pitted stimuli that could have been produced by the human vocal tract to stimuli that the human tract is incapable of producing. Thus, we ignore whether the infant's behaviour is determined by a speech vs. non-speech contrast or by a contrast between vocal vs. non-vocal-like sounds. Belin et al (2000)

have recently claimed that the human brain has an area that is devoted to processing conspecific vocal productions. Adults in an fMRI experiment listened to various
speech and non-speech sounds (laughs, coughs, sighs, etc.) generated by the human
vocal tract. The authors reported that sounds produced by the vocal tract elicit
greater activation than non-vocal sounds bilaterally in non-primary auditory cortex.
However, vocal sounds elicit greater activation than non-vocal sounds bilaterally
along the superior temporal sulcus (STS). On the basis of these results they argue
that there is a 'voice-region' much as there is a 'face-region'. Conceivably, under different experimental conditions, one could find areas that are selectively activated by
speech-like stimuli that the human vocal tract could generate, as compared to similar
stimuli that it could not generate. Does the brain have a localizer for processing
human voices much as Kanwisher et al (1997) have proposed that faces are
processed in the FFA (fusiform face area)? According to Belin and colleagues,
human voice is processed in the STS. This conclusion may be premature and more
experiments would be needed to be convincing.[4]

In our laboratory we have studied the specificity of the cortical areas devoted to
processing different information-types, before any learning has occurred. Establishing that a brain area is a localizers of a function does not tell us how the area
acquired this function. Our own efforts centre on the initial state of the cognitive
system and of the brain structures that support it. Adults have already learned how
to process and encode faces or human vocal tract productions, and might as a result
have cortical tissue dedicated to this competence. Therefore, in order to distinguish
aptitudes that arise as part of our endowment from those that arise as a consequence of learning, it is useful to investigate very young infants and, whenever possible, neonates. Indeed, during the first months of life, infants acquire many
language specific properties (see Werker & Tees 1984, Kuhl et al 1992, Mehler &
Dupoux 1994, Jusczyk 1997).

Standard neurological teaching tells us that the left hemisphere (LH) is more
involved in language representation and processing than the right hemisphere (RH)
(see Dronkers 1996, Geschwind 1970, Bryden 1982, among many others, but see

---

[4] To establish that the FFA is an area that is specifically responsive to faces, Kanwisher had to test
many other stimuli and conditions. Gauthier et al (2000) have challenged the existence of the
FFA showing that this area is also activated by other sets of stimuli whose members belong to
categorized ensembles even though they are not faces. Indeed, Gauthier and her colleagues
showed that when Ss learn a new set before the experiments, its members then activate the FFA.
Gauthier argued that her studies show that the FFA is not a structure uniquely devoted to face
processing. Without denying the validity of Gauthier's results, Kanwisher still thinks that the FFA
is a *bona fide* face area. We think that although we understand the FFA much better than Belin's
voice area we still have to be very careful before we accept the proposed locus as a voice-specific
area. *A fortiori* we need equal parsimony before we admit that we do have a specific voice-
processing area. Future research will clarify this issue.

also Gandour et al 2002). Are infants born with specific LH areas devoted to speech processing or is LH specialization solely the result of experience? The response to this question is still tentative. Some studies report that infants are born with speech processing abilities similar to those of experienced adults. For instance, infants discriminate all the phonetic contrasts that arise in natural languages (Jusczyk 1997, Mehler & Dupoux 1994). At first, this finding was construed as showing that humans are born with specific neural machinery devoted to speech. Subsequent investigations, however, demonstrated that basic acoustic processing capacities explain these early abilities that humans share with other organisms (Jusczyk 1997, Jusczyk et al 1977, Kuhl & Miller 1975). Thus, though it is conceivable that humans are endowed with a species-specific disposition to acquire natural language, we lack the data that might answer whether we are born with cortical structures specifically dedicated to the processing of speech.

Experimental psychologists devoted substantial efforts to establish whether LH superiority is the consequence of language acquisition or whether language is mastered because of this cortical specialization. Most studies have found an asymmetry in very young humans (Best et al 1982, Bertoncini et al 1989, Segalowitz & Chapman 1980). A few ERP studies have also found trends for LH superiority in young infants (Molfese & Molfese 1979, Dehaene-Lambertz & Dehaene 1994). Both the behavioural and the ERP data suggest that LH superiority exists in the infants' brain.

Below we review results obtained using more advanced imaging methods to study functional brain organization in newborn infants. Several methods are being pursued in parallel. A few groups have begun to study healthy infants using fMRI (Dehaene-Lambertz et al 2002). In the following section, we focus on recent results we obtained with Optical Topography (OT).

Brain specialization in newborns: evidence form OT

Optical Topography is a method derived from the Near Infrared Spectroscopy (NIRS) technology developed in the early 1950s (see Villringer & Chance 1997 for an excellent review of the field). This technology allows us to estimate the vascular response of the brain following stimulation.[5] In particular, it allows one

---

[5] This non-invasive device uses near-infrared light to evaluate how many photons are absorbed in a part of the brain cortex following stimulation. Like fMRI, it estimates the vascular response in a given area of the cortex. As fMRI, it estimates changes in deoxHb, however, it also gauges changes in oxyHb correlated with stimulation. Like fMRI, its time resolution is poorer than that of ERP. Our device uses bundles of source and detector fiber optics that are applied to the infants' head. The source fibers deliver near-infrared light at two wavelengths. One of the wavelengths is better absorbed by oxyHb, while the other is better absorbed by deoxyHb.

to estimate the concentration of oxy-haemoglobin (oxyHb), deoxy-haemoglobin (deoxyHB) and total haemoglobin over a given area of the brain.

Peña et al (2003) used a NIRS device to evaluate whether the neonate's brain is specifically tuned to human speech. Using sets of light emitting fibres and light detecting fibres, and two wavelengths, one can observe how the cortex responds to stimuli on homologous areas of the LH and RH. We placed the probes so as to measure activity over the RH and LH temporal and parietal areas. Participants were tested with Forward Speech (FW): infants heard sequences of 15 seconds of connected French utterances separated from one another by periods of silence of variable duration, i.e. from 25–35 s. In another condition, Backward Speech (BW), infants were tested as in the FW condition but with the speech sequences played backwards. In this second condition the speech signal was converted from FW to BW using a speech waveform editor. Ten such blocks of FW and BW conditions were used. Finally, in a control condition, infants were studied in total silence for a duration identical to that of the above conditions.

The results show that, as in adults, the haemodynamic response begins four to five seconds after the infant receives auditory stimulation. This time-locked response appears more clearly for the oxyHB than for the deoxyHB.[6] The results also show that roughly five seconds after the presentation of the FW utterances, a robust change in the concentration of totalHB takes place over the temporo-parietal region of the LH. Interestingly, the concentration of totalHb is relatively lower and comparable both in the BW and in the Silence conditions. Thus only forward speech gives rise to a significant increase in total Hb over the LH, while BW speech does not give rise to a significant increase in total HB in any of the channels. While the acoustic energy is identical in the FW and BW sentences, and their spectral properties are mirror images of each other, the brain responds very differently to the acoustic pattern that can be produced by the human vocal tract in contrast to that which cannot.

The reported results suggest that the brain of the newborn infant responds differently to natural and backward speech. To understand the singularity of this result, it may be useful to mention that in a pilot study we found that monolingual adults who were tested with materials similar to those used with infants are sometimes tricked to believe that both FW and BW are sentences in some foreign languages. Interestingly, if they are asked to rate which one sounds more 'natural', they tend to choose forward speech. The BW and FW utterances are indeed very similar. FW and BW speech differ in terms of their timing patterns. Indeed, final lengthening appears to be a universal property of natural language. Thus, only BW utterances

---

[6]We now know that a better choice of wavelengths would have permitted us to avoid this problem.

have initial lengthening. In addition, some segments, (stops i.e. [p], [t], [k], [b], [d] and [g] and affricates, like [ts] or [dz]), become very different when played backwards. The vocal tract cannot produce backward speech. Since infants cannot produce forward speech either, they might have ignored the contrast between the BW and FW conditions (Liberman & Mattingly 1985). However, since the neonate's brain responds differently to FW and BW we infer that, in some sense, the infants' brain has become attuned to the difference between natural and unnatural utterances. We might tentatively attribute this result to the specialization of certain cortical areas of the neonate's brain for speech. Humans might have, like many other vertebrates, specialized effectors and receptors for a species-specific vocalization, which in our case is speech. This possibility needs to be studied in greater detail. We are replicating this result using an improved NIRS device. The new machine has wavelengths that are better suited to track vascular responses and moreover it is equipped with probes that are designed to fit better on the infant head.

It is not only necessary to replicate the above result but it is also necessary to have a better theoretical grasp of what these results entail. If the LH dominance already observed in neonates is viewed as an emergent evolutionary module then we ought to explore whether asymmetrical patterns of activation to FW and BW speech are also found in non-human primates or even more primitive vertebrates. As a matter of fact, work with monkeys exists and suggests that their behaviour is similar to that of infants, when exposed to FW and BW speech. In a series of studies comparing the human newborn and the adult cotton-top tamarin monkey for their behavioural responses to FW and BW speech, Ramus et al (2000) showed that, like infants, tamarins discriminate two different languages (Japanese and Dutch) when the utterances are played forwards but fail to do so when the utterances are played backwards. The ability to behave like infants in the FW and BW conditions is remarkable since tamarins will never develop speech. This outcome ought to temper the desire to conclude that the infant results are based on a species-specific system to process natural speech. Indeed, the observed specialization may have arisen much before language arose. Many vertebrates produce cries and vocal noises in this way and a specialized module might have evolved to discriminate such sounds from other sounds that cannot be generated by these kinds of vocal tracts.

Obviously, the advent of imaging studies using neonates will permit more precise investigations to establish whether the specialization for speech is really present at birth or whether there is activation for streams of sounds that can be produced by the vocal tract of any vertebrate species. In the meantime, these studies have shed some light on complex issues that were hard to study with more traditional behavioural methods. To close this section let us just remind the reader of a study carried

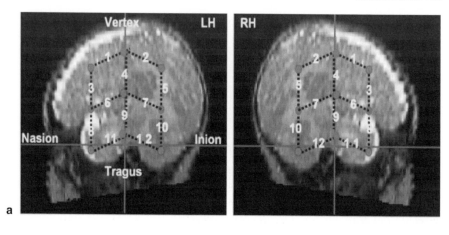

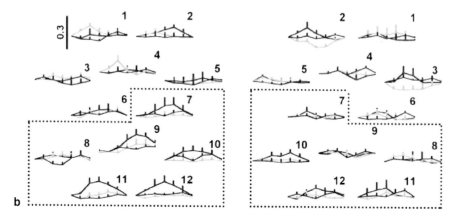

FIG. 1.   Changes in total haemoglobin for newborn Italian infants. Each infant contributes more than three blocks in each one of the conditions. All blocks are summed across infants. (a) Indicates how the probes were placed over the left and right hemispheres (LH, RH). (b) Results showing the activity recorded over each one of the hemispheres. Darkest grey, forward speech; lightest grey, backward speech; intermediate grey, silence. Reproduced by permission of Peña et al (2003).

out with three-month-olds tested with FW, BW and Silence using an fMRI device. Dehaene-Lambertz et al (2002) showed that cortical regions were active well before the infant has acquired the native language.

Let us now turn to another property that is essential to understand the first adaptations that the human infant makes to speech stimuli, that is rhythm. Not only did linguists use the notion of rhythm to sort languages into groups or classes but, independently, developmental psychologists discovered that rhythm is the very first adjustment that infants make to the maternal language.

Neonates use rhythm to tune into language

The notion of rhythm relates to the relative duration of constituents in a sequence. What, we can ask, are the elements responsible for the characteristic perception of the rhythm of a language? Three constituents, or atoms, of different size have been proposed to be roughly isochronous in different languages, thus giving rise to rhythm: syllables, feet and morae (Pike 1945, Abercrombie 1967, Ladefoged 1975). Syllables have independently been construed as a basic constituent or atom for both speech production and speech comprehension (Levelt 1989, Mehler 1981, Cutler & Mehler 1983). Infants begin to produce syllables several months after birth, with the onset of babbling. However, the infant may rely on syllables to process speech before s/he produces syllables. If so, it should be possible to find indications that neonates process syllables in linguistic-like ways.[7] Bertoncini & Mehler (1981) explored this issue using the non-nutritive sucking technique and showed that very young infants distinguish a pair of syllables that differ only in the serial order of their constituent segments, e.g. *PAT* and *TAP*. The infants, however, failed to distinguish a pair of items, i.e. *TSP* and *PST*, which were derived from the previous ones by replacing the vowel [a] by the consonant [s]. This editing of the 'good' syllables transforms the new items into 'marked' or 'bad' syllables. To understand the infants' failure to distinguish this pair, we ran a control experiment, in which infants were presented with the same 'marked' syllables inserted in a context of vowels, i.e. *UPSTU* and *UTSPU*, that generated two bi-syllabic, well-formed speech sounds. When these sequences were presented to the infants, discrimination ability was restored. This experiment suggests that the infant discriminates items presented in a linguistic-like context but s/he neglects those constructs in other acoustic contexts.

---

[7] A universal property of syllables is that they have an obligatory *nucleus* optionally preceded by an *onset* and followed by a *coda*. While onset and coda positions are occupied by consonants (C), the nucleus is generally occupied by a vowel (V). In some languages, the nucleus can be occupied by a sonorant consonant, in particular [r] and [l]. Thus, a syllable may not contain more than one vowel (or a dipthong). CV is the optimal syllable, i.e. the onset is present and the coda absent. All natural languages have CV syllables. There is a hierarchy of increasing complexity in the inclusion of syllable types in a given language. Thus, a language that has V will also have CV, but not vice versa. A language that has V, instead, does not necessarily have VC. That is, in some languages all syllables end in a vowel. Similarly, a language that has CVC will also have a CV in its repertoire. A language that includes a CCV in its repertoire will have CV and a language that includes CVCC also has CVC. The prediction then is that while CVC is a well-formed potential syllable in many languages, CCC is not, especially if none of the consonants is sonorant.

As we mentioned in footnote 5, some languages, (e.g. Croatian and some varieties of Berber) allow specific consonants to occupy the nuclear position of the syllable. For instance, in Croatian, *Trieste*, the Italian city, is called *Trst* where [r] is the nucleus. This is not an exceptional case in the language. Indeed, the word for 'finger' is *prst*, the word for 'pitcher' is *vrča*, and 'square' or 'piazza' is *trg*. Why then did the infants respond as they did in the results reported in Bertoncini & Mehler (1981)? Why did the infants fail to treat *PST* and *TSP* as different syllables? One explanation may be that we tested rather old (i.e. two-month-olds) infants who had already gained considerable experience about the surrounding language. All the infants had been tested in a French environment; it is possible that the stimuli were already considered inappropriate for their language and thus they neglected the ill-formed stimuli. An alternative explanation that still needs to be explored is that PST and TSP are non-standard syllables in any language, including the ones named above. To the best of our knowledge there is no language that allows [s] as a syllabic nucleus. We predict that infants have no difficulty in distinguishing pairs in which [r] or [l] figure as nuclei (e.g. [prt] vs. [trp] or [plt] vs. [tlp]) since such syllables occur in more than a few languages, but that they will have difficulty distinguishing sequences in which the nuclear position is occupied by [s] or [f], (e.g. [pst] vs. [tsp] or [pft] vs. [tfp]). To ensure that the infant has not become familiar with the syllable repertoire in the surrounding language, we are testing neonates in their first week of life.

Bijeljac-Babic et al (1993) had already claimed that very young French raised infants attend to speech using syllabic units; that is, units that are related to the rhythmical pattern instantiated in that language. These authors showed that infants distinguish lists of bisyllabic items from a list of trisyllabic ones. They used CVCV items (e.g. *maki, nepo, suta, jaco*) and CVCVCV items (e.g. *makine, posuta, jacoli*). This result is observed regardless of whether the items differ or are matched for duration. Indeed, some of the original items were compressed and others expanded to match the mean durations of the two lists. Infants discriminated the lists equally well, suggesting that either the number of syllables or just the number of vowels is what counts. We focused on syllables rather than on feet or morae because of the total absence of studies that explored whether neonates can also represent those units. Below we will explain why we believe that syllables, or possibly vowels, play such an important role during the early steps of language acquisition.

The results described above fit well with recent evidence showing that neonates are born with remarkable abilities to learn language. For instance, in the last decade numerous studies have uncovered the exceptional abilities of babies to process the prosodic features of utterances (Moon et al 1993, Mehler et al 1988). Indeed, for many pairs of languages, infants tend to notice when a speaker switches from one language to another. What is the actual cue that allows infants to detect this switch?

The essential property appears to be linguistic rhythm, defined as the proportion in the utterances of a language that is occupied by vowels (Ramus et al 1999). If two languages have different rhythms (an important change in %V) the baby will detect a switch from one language to the other. If languages have similar rhythms, as for instance, English and Dutch or Spanish and Italian, very young infants will fail to react to a switch (Nazzi et al 1998).

The variability of the intervocalic interval (i.e. ΔC, the standard deviation of the intervocalic intervals) also plays an important role in explaining the infants' behaviour. In fact, ΔC in conjunction with %V provides an excellent measure of language rhythm that fits well with the intuitive classification of languages that phonologists have provided. Indeed, their claim is that there are basically three kinds of rhythm depending on which of the three possible units maintains isochrony in the speech stream: stress-timed rhythm, syllable-timed rhythm and mora-timed rhythm (Pike 1945, Abercrombie 1967, Ladefoged 1975). However, once exact measures were carried out, isochronous units were not found (see Dauer 1983, Manrique & Signorini 1983, but see Port et al 1987 who have claimed that there is a fair amount of isochrony for morae). This does not mean, as one might have argued, that the classification linguists proposed on the basis of their intuitions has to be dismissed. Rather, Ramus et al (1999)'s definition of rhythm on the basis of ΔC and %V divides languages exactly into those three intuitive classes, as shown in Figure 2.

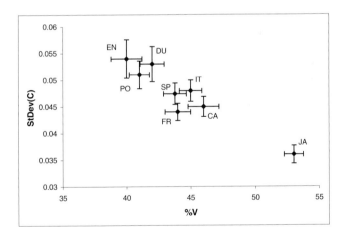

FIG. 2. %V is the mean proportion of the utterances in a language that is occupied by vowels and ΔC or StDev(C) is the standard deviation of the consonantal intervals. The plot incorporates eight languages spoken by four female speakers. Each speaker utters 20 sentences (each language is represented by 20 utterances). The distribution of the languages is compatible with the notion that they can be grouped into three classes as predicted by linguists' intuitions. Reprinted from Ramus et al (1999), with permission from Elsevier.

A language with a high %V and a low ΔC (like Japanese or Hawaiian) is likely to have a small syllabic repertoire. Mostly, such languages allow only CVs, and Vs, giving rise to the typical rhythm of the mora-class. Moreover, intervocalic intervals cannot be very variable since consonant clusters are avoided, and codas are in general disallowed. In Japanese, for instance, codas generally contain /n/ (as in the word *Honda*).[8] Romance languages, as depicted in Figure 2, have a smaller value of %V because their syllabic repertoires are larger. Indeed, these languages allow both onsets and codas. Moreover, onsets may contain consonant clusters (e.g. *prêt*, *prato*) and, at least in some Romance languages, even codas contain more than one consonant (e.g. *tact*, *parc*). However, fewer syllable types are allowed in Romance languages than in stress-timed languages like Dutch or English. Indeed, while in Romance languages the typical syllabic repertoire ranges from six to eight syllables, Germanic languages have over sixteen syllable types. This conception of rhythm relates to Dauer (1983) and also Nespor (1990) who claim that linguistic rhythm is a side effect of the syllabic repertoire that languages instantiate. Languages such as Japanese have a very restricted syllable repertoire, and thus a relatively high proportion of utterances is taken up by vowels. In contrast, languages with a large number of syllable types, and thus many consonant clusters, tend to have a smaller proportion of the utterances taken up by vowels. Interestingly, one could conclude that if a larger number of languages were included in Figure 2, it might turn out that some more classes or even a continuum is obtained rather than the clustering of languages into the few classes that we now observe. However, if the notion of rhythm is really related to the claim according to which the number of syllable types is what gives rise to the intuitive notion of linguistic rhythm, things will go in favour of a clustering. Indeed, the syllable repertoires come in groups. Up until now, we have considered languages that have two or three syllable types (Hawaiian, Japanese, etc.), six to ten syllable types (Spanish, Greek, Italian, etc.) and languages that have sixteen or more (English, Dutch, etc.), see Nespor (1990). Future scrutiny with a larger set of languages will determine whether the notion that languages fall into a restricted number of classes is borne out or not, and if so, how many classes there are.

The conjecture we make is that rhythm, as defined by Ramus et al (1999), is sufficient to explain all the behavioural results showing that languages cluster into a few classes. Indeed, Ramus (1999) simulated the ability to discriminate switches from one language to another in infants and adults. He showed that %V is sufficient to account for all the available empirical findings involving neonates. This outcome sustains our resolve to pursue this line of investigation. Indeed, it seems

---

[8] Or geminates as in the word *Sapporo*.

unlikely that linguistic rhythm would be so salient for the neonate without having any further impact on how language is learned.

The first known adjustment to the surrounding language the neonate makes concerns rhythm. The processing of linguistic rhythm changes over the first two months of life. Mehler et al (1988) remarked that while American two-month-olds fail to discriminate Russian from French, one-week-old French infants successfully discriminate not only Russian from French but also English from Italian suggesting that by two months of age infants have encoded some properties of their native language and stop discriminating between two unfamiliar rhythms. Such a bias may explain the observed failure to discriminate a switch between two 'unknown' languages. Christophe & Morton (1998) further investigated this issue, testing two-month-old and four-month-old British infants. They found that the infants were able to discriminate a switch between English and Japanese but not a switch between French and Japanese. Presumably, the former pair of languages is discriminated because it involves one familiar and one novel type of rhythm. The second switch is not discriminated because neither language has a rhythm that is familiar to the infant. To buttress their interpretation, Christophe & Morton (1998) also tested the behaviour of the same British infants with Dutch. First, they corroborated their prediction that these infants would fail to discriminate Dutch from English, because the two languages have a similar rhythm. Next, they showed that the infants discriminate Dutch from Japanese, two languages foreign to these infants. In fact, while Dutch differs from English, their rhythm is similar, and thus, although Dutch is not their native language it still catches the infants' attention.

We hope to complement the above behavioural research with brain-imaging methods. If we succeed, this will provide more information to decide whether learning and development of language require a passage through an attention-drawing device based on rhythm. But even before we obtain such data we have to raise the following central question: Why are infants interested in rhythm even before the elementary sound patterns of utterances attract their attention?[9] What information does linguistic rhythm provide to render it so relevant for language acquisition? We have implemented two procedures to answer these questions. First, we have tried to gather data using optical topography to pursue the exploration of language processing in the neonate, as described above. Second, we have explored the potential role of rhythm in other areas of language acquisition. Specifically, we asked whether rhythm might play a role in the setting of syntactic parameters, and also whether it might be exploited in segmentation, as described in the following sections, see also Mehler & Nespor (2004).

---

[9] Werker & Tees (1983) were the first to point out that the first adjustment to the segmental repertoire of the language of exposure becomes apparent at the end of the first year of life.

Segmenting the speech stream

Ramus (1999) conjectured that language rhythm provides the infant with information about the richness of the syllabic repertoire of the language of exposure (cf. Dauer 1983 and Nespor 1991). For the sake of argument, we assume that the infant gains this type of information from rhythmic properties in the signal. What could the use of such information for the language-learning infant be? What benefit could a baby draw once s/he learns that the number of syllable types is four, six or 16? Will such information help perception of speech? Or will such information be essential in mastering the production routines or elementary speech acts? We cannot answer these questions in detail. However, there is no reason to believe that knowing the size of the syllabic repertoire facilitates perception of speech. Is there evidence that a learner performs better when s/he has prior knowledge of the number of types or items in the set to be learned? We can give an indirect answer to this question by looking at lexical acquisition. Infants appear to learn the lexicon without ever knowing or caring whether they have to master 4000 or 40 000 words. Why would knowledge of the number of syllable types be necessary given that infants acquire thousands of words without, to the best of our knowledge, requiring special signals about word types? However, there is a tentative explanation for the infant's precocious interest in rhythm. Rhythmic information may constrain lexical acquisition. Indeed, the size of the syllabic repertoire is inversely correlated with the mean length of words. Hence, gaining information about rhythm may provide a bias to look for large or smaller lexical items in the language of exposure (Nespor & Mehler 2003). This simple procedure may prove important to understand how infants identify potential words since we know that most words come in packages of fluent, continuous streams.

It is too early to decide whether infants make use of the above bias. Many other conjectures have been proposed. For instance, it has been suggested that many words are first learned when the mother produces them in isolation. Others have suggested that infants focus on the onset and the end of all the utterances they receive. This might allow them to isolate recurrent items. None of these conjectures should be ruled out. However, there is a proposal that infants parse continuous streams by finding dips in transition probabilities between some syllables and high transition probabilities between other syllables. (Saffran et al 1996).

Can rhythm also help segmenting the continuous speech stream? Nespor et al (2003) have proposed that infants who listen to a language with a %V that is higher than 50%, like in 'mora-timed' languages, will tend to parse signals looking for long word-like constituents while infants who listen to a language whose %V is below 40% will tend to search for shorter units. This follows from the fact that the syllabic repertoire in, e.g. Japanese, is very limited, which entails that monosyllables will be rare and long words will be very frequent. We assume that speakers are unwilling to put up with polysemy to an extent that would threaten communication. Supposedly, languages are designed to favour rather than to hinder communication. In fact, words turn out to be long in Japanese as well as in any other language with a restricted syllabic repertoire. In contrast, languages such as Dutch or English, which have a very rich syllabic repertoire (%V close to 45%), allow for a large number of different syllable types. Hence, it is easy to understand why among the first 1000 words in the language many will be monosyllables (nearly 600 out of 1000). Languages like Italian, Spanish or Catalan, whose %V lies between that of Japanese and English, also have an intermediate number of syllable types. As expected, the length of the most common words falls between two and three syllables.

Assuming that rhythmic properties are important during language acquisition and, furthermore, that very young infants extract the characteristic rhythm of the language of exposure, it would be useful to understand the underlying computational processes. Unfortunately, at this time, we have no results that might allow us to explain how these computations are performed. Hopefully, future studies will clarify whether the auditory system is organized to extract rapidly and efficiently the rhythmic properties of the speech stream, and/or whether we are born to be powerful statistical machines that allow for small differences in rhythm between classes of languages to be detected. Regardless of how the properties that characterize rhythmic classes are identified, our conjecture is that the trigger that leads the infant to expect words of a certain length is determined by rhythm. Once rhythm has set or fixed this bias, one may find that infants segment speech relying on other mechanisms. For example, the statistical computations that Saffran and her colleagues have invoked (see below) may be an excellent tool to segment streams of speech into constituents. Regardless of the putative role of rhythm we acknowledge that statistical process are powerful and well attested as an instrument for segmentation while rhythm is only very indirectly related to segmentation: it is relevant for the infants and it predicts the mean length of words in languages according to their syllabic structure.

Saffran et al (1996) and Morgan & Saffran (1995) have revived the view that statistical information plays a central role in language acquisition. Indeed, Miller (1951) had already postulated that the statistical properties of language could help process signals and thus favour language acquisition. Connectionism has also highlighted the importance of statistics for language learning; it postulates that the language learner can be viewed as a powerful statistical machine. We acknowledge that the advantage of statistics is that it can be universally applied to unknown languages, and thus pre-linguistic infants may also exploit it.

Saffran et al (1996) have shown that adults and nine-month-old infants confronted with unfamiliar monotonous artificial speech streams tend to infer word boundaries through statistical regularities in the signal. A word boundary is postulated in positions where the transitional probability (TP)[10] drops between one syllable and the next.[11] Participants familiarized with a monotonous stream of artificial speech recognize trisyllabic items delimited by dips in TP. As an example, imagine that *puliko* and *meluti* are items with high TPs between the constituent syllables. If participants are asked which of *puliko* or *likome* (where *liko* are the last two syllables of the first word and *me* the first syllable of the second word) is more familiar, they select the first well above chance. Among a large number of investigations that have validated Saffran et al's findings, we have found that, by and large, French and Italian adult speakers perform as the English speakers of the original experiment.[12]

Let us summarize what we have tried to suggest this far. We have noticed that linguistic rhythm can be captured as suggested by Ramus et al (1999) by measuring the amount of time/utterance occupied by vowels and by the variability of intervocalic intervals. We also acknowledged the powerful role that statistics plays in helping determine early properties present in the speech stream. However, we also noticed that rhythm as defined in Ramus et al presupposes that our processing system makes a categorical distinction between consonants and vowels. In the following section we expand on the notion that there is a basic categorical distinction between Vs and Cs and we go on to propose a view of language acquisition based on the consequences of this divide.

Rhythm, signals and triggers

Developmental psycholinguists and students of adult language perception and production have tried to evaluate whether the rhythmic class to which a language belongs is related to phonological units that are highlighted during processing, see Cutler (1993). More recently, linguists and psycholinguists have started exploring whether phonological properties related to syntax can guide the infant in the setting of the essential parameters necessary to acquire the grammar of the language. We

---

[10] Transition probability between two syllables is synonymous with the conditional probability that the second syllable will occur immediately after the first one.

[11] Saffran, Aslin & Newport (1996) use streams that consist of artificial CV syllables that are assembled without pauses between one CV and the next. All syllables have the same duration, loudness and pitch. TPs between adjacent syllables (within trisyllables) range from 0.25 to 1.00. The last syllable of an item and the first syllable of the next one have TPs ranging from 0.05 to 0.60.

[12] One divergence between the results reported by the Rochester group and our own concerns the computation of TPs on the consonantal and vocalic tiers. Native English speakers can use both tiers to calculate TPs (Newport & Aslin 2004). Our own Ss, regardless of whether they are native French or native Italian speakers, can only use the consonantal tier. Notice that Newport & Aslin use only two families, generating repetitions that we did not allow (see main text below).

are presently exploring to what extent linguistic rhythm relates to the processes that leads to the discovery of the non-universal properties of his/her native syntax.

Our proposal is to integrate P&P with a general theory of learning. While it is commonly taken for granted that general learning mechanisms play a role in the acquisition of the lexicon (Bloom 2000), their role in the actual setting of parameters has not been sufficiently explored. In fact, while signals may give a cue to the value of a certain parameter, general learning mechanisms may play a role in establishing the validity of such a cue. For instance, in order to decide whether complements in a language precede or follow their head, it is necessary to establish whether the main prominence of its phonological phrases is rightmost or leftmost, as we will see below. Within a language, syntactic phrases are, by and large, of one type or another, i.e. they are either Head-Complement (HC) or Complement-Head (CH).[13] There are languages, however, in which the word order in a specific phrase can be different from the standard word order. Since the pre-lexical infant ignores whether exceptions of this kind weaken the information that overall prominence provides, there must be some mechanism for her/him to detect such cases. In all likelihood, statistical computations allow the infant to discover and validate the most frequently used phonological pattern that can act as a cue to the underlying syntax (Nespor et al 1996). Indeed, even an infant who is exposed to a regular language (as to the HC order) may occasionally hear irregular patterns, e.g. foreign locutions or speech errors. In this case, the frequency distribution difference between the occasional and the habitual patterns will allow the infant to converge on the adequate setting.

Let us focus more closely on the case of the HC parameter. In the great majority of languages, the setting of this parameter simultaneously specifies the relative order of heads and complements and thus of main clauses with respect to subordinate clauses. That children start the two-word stage without making mistakes in word order suggests that this parameter is set precociously (Bloom 1970, Meisel 1992). In addition, even prior to this, babies react differently to the appropriate as compared to the wrong word order (Hirsh-Pasek & Golinkoff 1996). These facts suggest that children must set this parameter quite early in life. Given such evidence a scenario in which the infant finds how to set basic parameters prior to, or at least independently of the segmentation of the speech stream into words seems sensible to explore. If the child sets parameters before learning the meaning of words,

---

[13] For example, in languages in which the verb precedes the object, subordinate clauses follow main clauses. In contrast, in languages in which the verb follows the object, subordinate clauses precede the main clause. Other ordinal properties also correlate with the HC or CH structure of languages; for a more technical definition of the notion of the head-complement parameter see, e.g. Haegeman (1994).

prosodic bootstrapping might become immune to the paradox pointed out by Mazuka (1996). She observes that to understand the word order of, say, heads and complements in the language of exposure, an infant must first recognize which is the head and which is the complement. But once the infant has learned to recognize which word in a pair of words functions as the head and which as the complement, it already knows how they are ordered. If you know how they are ordered, the parameter becomes pointless for the purposes of acquisition. Without syntactic knowledge, word meaning cannot be learned and without meaning, syntax cannot be acquired either.

How can a child overcome this quandary and get information about word order just by listening to the signal? What is there in the speech stream that might provide a cue to the value of this parameter? Rhythm, in language as in music, is hierarchical in nature (Liberman & Prince 1977, Selkirk 1984). We have seen above that at the basic level, rhythm can be defined on the basis of %V and ΔC. At higher levels, the relative prominence of certain syllables (or the vowels that form their nuclei) with respect to other syllables reflects some aspects of syntax. In particular, in the phonological phrase,[14] rightmost main prominence is characteristic of HC languages (such as English, Italian or Croatian) while leftmost main prominence characterizes CH languages (such as Turkish, Japanese or Basque) (Nespor & Vogel 1986). A speech stream is thus an alternation of words in either weak–strong or strong–weak chunks. Suppose that this correlation between the location of main prominence within phonological phrases and the value of the HC parameter is indeed universal. Then we can assume that by hearing either a weak–strong or a strong–weak pattern, an infant becomes biased to set the parameter to the correct value for the language of exposure. The advantage of such a direct connection between signal and syntax (Morgan & Demuth 1996), is that the only prerequisite is that infants hear the relevant alternation. To see whether this is the case, Christophe et al (2003) carried out a discrimination task using resynthesized utterances drawn from French and Turkish sentences. These languages have similar syllabic structures and word-final stress but they differ in the locus of the main prominence within the phonological phrase, an aspect that is crucial for us.[15] The experiment used delexicalized sentences pronounced by the same voice.[16] Six to 12-week old infants discriminated French from Turkish. It was concluded that infants discriminate the two languages only on the basis of the different location of the

---

[14] The phonological phrase is a constituent of the phonological hierarchy that includes the head of a phrase and all its function words, e.g. articles, prepositions and conjunctions; for a more technical definition see Nespor & Vogel (1986).
[15] The effect of the resynthesis is that all segmental differences are eliminated.
[16] Sentences were synthesized using Dutch diphones with the same voice.

main phonological phrase prominence. Knowing that infants discriminate these two types of rhythmic patterns opens a new direction of research to assess whether infants actually use this information to set the relevant syntactic parameter.

The C/V distinction and language acquisition

Why does language need to have both vowels and consonants? According to Plato rhythm is 'order in movement'. But why, at one level of the rhythmic architecture, is the order established by the alternation of vowels and consonants? Why do all languages have both Cs and Vs? Possibly, as phoneticians and acousticians argue, see Stevens (1998), this design structure has functional properties that are essential for communication. Indeed, vowels have considerable energy, allowing them to carry the signal, while consonants are modulations allowing for an increase in the number of messages with different meanings that can be transmitted. Even if this explanation is correct, the reason languages necessarily include both vowels and consonants may have additional functional roles. Indeed, Nespor et al (2003) have proposed that vowels and consonants play a different functional role in language acquisition and language perception. Consonants are intimately linked to the lexicon structure, while vowels are linked to grammatical structures.

The lexicon allows the identification of thousands of lemmas, while grammar organizes the lexical items in a regular system. There is abundant evidence that consonants are more distinctive than vowels. For instance, cross-linguistically there is a clear tendency for Cs to outnumber Vs: the most frequent segmental system in the languages of the world has five vowels and around 20 consonants. But languages with just three vowels are also attested and historical linguists working on common ancestors of different languages have posited two or even one single vowel for proto-Indo-European. However, languages attested today have at least two vowels. For example, the Tshwizhyi and Abzhui dialects of Abkhaz contrasts only /a/ and /i/, with significant allophony.[17]

A widespread phenomenon in the languages of the world is vowel reduction in unstressed positions. Languages like English, in which unstressed vowels are centralized to schwa, thereby losing their distinctive power, represent an extreme case. Another widespread phenomenon is vowel harmony, whereby all the vowels in a certain domain share some features. No comparable phenomena affect consonants. The pronunciation of Cs is also less variable (thus more distinctive) than that of Vs. Prosody is responsible for the variability of vowels within a system: both rhythmic and intonational information (be it grammatical or emotional) is by and large

---

[17] Some linguists claim that it is possible to posit only *one* vowel in some Abkhaz dialects, though the general consensus seems to be that that is stretching things a bit.

carried by vowels. Acoustic-phonetic studies have documented that while the production of vowels is rather variable, consonants are more stable. Moreover, experimental studies have shown that while consonants tend to be perceived categorically, vowels do not (see Kuhl et al 1992, Werker & Tees 1984). These different reasons for the variability of vowels, of course, make them less distinctive. Evidence for the distinctive role of consonants is also attested by the existence of languages (e.g. Semitic languages) in which lexical roots are composed uniquely of consonants. To the best of our knowledge, there is no language in which lexical roots are composed just of vowels.

The above noted asymmetry between Vs and Cs in linguistic systems is reflected in language acquisition. The first adjustments infants make to the native language are related to vowels rather than to consonants. Indeed, several pieces of evidence can be advanced to buttress this assertion. Bertoncini et al (1988) showed that neonates presented with four syllables in random order during familiarization react when a new syllable is introduced, provided that it differs from the others by at least its vowel. If the new syllable differs from the other syllables only by the consonant, its addition will be neglected.[18] However, two-month-olds show a response to both, i.e. whether one adds a syllable that differs from a member of the habituation set by its vowel or by its consonant. We must remember, however, that the above results are not due to limitations in discrimination ability but rather to the way in which the stimuli are represented. We can conclude that the first representation privileges vowels, but that by two months of age vowels and consonants are sufficiently well encoded as to yield a similar phonological representation. Similarly, while by six months of age infants respond preferentially to the vowels of their native language,[19] Werker & Tees (1984) have shown that convergence to native consonants happens later: consonantal contrasts that are not used in the native language are still discriminated before eight months and are neglected only a few months later. That is, when the infant goes from phonetic to phonological representations, vowels seem to be adjusted to the native values before consonants. This observation is yet another indication that vowels and consonants are categorically distinct from the onset of language acquisition. Our suggestion is that these two categories have a different function in language and in its acquisition.

As we mention below, vowels and consonants, even when they are equally informative from a statistical point of view, are not exploited in similar ways.

---

[18]Two kinds of habituation were used, [bi], [si], [li] and [mi] or [bo], [bae], [ba] and [bo]. The introduction of [bu] causes the neonate to react to the modification regardless of the habituation. The introduction of [di] after the neonate is habituated with the first set of syllables is neglected and so is the introduction of [da] after habituation with the second set.

[19]American infants respond preferentially to American vowels as compared to Swedish vowels while Swedish infants respond preferentially to Swedish vowels compared with English ones (Kuhl et al 1992).

Newport & Aslin (2004) used a stream of synthetic speech consisting of CV syllables of equal pitch and duration in which 'words' are characterized only by high TPs between the consonants, while vowels change in the different instantiations of a 'word'. The authors showed that participants have no difficulty segmenting such streams.[20] We replicated this robust finding with Italian and French-speaking participants (Peña 2002, Bonatti et al 2005). In a similar experiment in which the statistical dependences were carried by vowels while the intervening consonants vary, the participants tested at Rochester were able to segment the streams while our participants (French, Spanish or Italian native speakers for the different experiments) failed to segment the stream into constituent 'words'. There are several differences between the English language experiment and the ones run using Italian or French. First, the streams Newport and Aslin used to test the vowel tier and the consonantal tier were not comparable to the ones used in Bonatti et al (2005). As we pointed out before, they used only two families to carry out their experiments while we used three. This means that they were obliged to repeat families while we carefully avoided such repetitions. The repetition of families might allow a repetition detection mechanism to intervene, see Endress et al (2005). This, of course, would only show that repetition detection promotes segmentation and not that the statistical dependencies incorporated in the vowel tier are responsible for the behaviour. If we are right, participants can use statistics to segment streams on the basis of the consonantal tier but not on the basis of the vowel tier.

On the basis of our experiments we conclude that a pre-lexical infant (or an adult listening to an unknown language) identifies word candidates on the basis of TP dips between either syllables or consonants, but not between vowels. However, we can ask why this should be so. As pointed out above, consonants change little when the word is pronounced in different emotional or emphatic contexts while vowels change a lot. Moreover, a great number of languages introduce changes in the vowels that compose a group of morphologically related words, i.e., *foot-feet* in English, and more conspicuously, in Arabic: *kitab* 'book', *kutub* 'books', *akteb* 'to write'. In brief, consonants rather than vowels are mainly geared to ensure lexical functions. Vowels, however, have an important role when one attempts to establish grammatical properties. We argued above that the rhythmic class of the first language of exposure is identified on the basis of the proportion of time taken up by vowels. Identification of the rhythmic class, we argued, provides information about the syllable repertoires, i.e. a part of phonology. Moreover, it gives information

---

[20] Thus, if a word has the syllables C-, C'-, C″ with the consonants that predict the next one exactly, regardless of the vowels that appear between them, the word in question will be preferred to a part word like C″-,C*-, C** (where stars illustrate that the two last syllables come from another 'word'). Of course, words have no probability dip between the consonants but part words enclose a TP dip between C″ and C*.

about the mean length of words in the language (see Nespor et al 2004). In addition, the information carried by vowels relates to the location of the main prominence within the phonological phrase. As was argued above, this prominence is related to a basic syntactic parameter.

Conclusion

In this paper, we have argued that both innate linguistic structure and general learning mechanisms are essential to our understanding of the acquisition of natural language. Theoretical linguists have focused their attention on the universal principles or constraints that delimit the nature of our endowment for language. Psychologists have explored how the child acquires the language of exposure, without showing much concern for the biological underpinnings of this process. After scrutinizing the potential limitations of both positions, we have pleaded in favour of the integration of the two approaches to improve our understanding of language acquisition.

Currently, there is a growing consensus that biologically realistic models have to be elaborated in order to begin understanding the uniqueness of the human mind and, in particular, of language.

In our research, we highlight the importance of the teaching of formal linguistics and explore how signals relate to the fixation of parameters. We have tried to demonstrate that signals often contain information that is related to unsuspected properties of the computational system. This has also obliged us to explore how signals can drive rule-like computations, see Peña et al (2002). We also laid out a proposal of how rhythm can guide the learner towards the basic properties of the language's phonology and syntax. In addition, we have argued that basic phonological categories, namely vowels and consonants, play different computational roles during language acquisition. These categories play distinctive roles across languages and appear to be sufficiently general for us to conjecture that they are a part of the species' endowment.

Another aspect that we highlight concerns the attested acoustic capacity of non-human vertebrates to discriminate and learn phonetic distinctions (see Kluender et al 1998, Ramus et al 2000). They also have the ability to extract and use the statistical properties of the stimulating sequences in order to analyse and parse them into constituents (M. Hauser, personal communication). These results suggest that humans and other higher vertebrates can process signals in much the same way. However, the fact remains that only humans, and no other animals, acquire the language spoken in the surrounds. Moreover, simple exposure is all that is needed for the learning process to be activated. Thus, we must search for the prerequisites of language acquisition in the knowledge inscribed in our endowment.

The fact that cues contained in the speech stream directly signal non-universal syntactic properties of language makes it clear that to understand how the infant attains knowledge of syntax, precociously and in an effortless fashion, attention must be paid to the cues that the signals provide. How can this argument be sustained when we have just acknowledged that human and non-human vertebrates processes acoustic signals in a similar fashion? The reason is that a theory of language acquisition requires not only an understanding of signal processing abilities, but also of how these cues affect the innate linguistic endowment. The nature of the language endowment, once precisely established, will guide us towards an understanding of the biological foundation of language, and thus will clarify why we diverge so significantly from other primates. This in turn, will hopefully lead us to formulate a testable hypothesis about the origin and evolution of natural language.

Acknowledgements

The research presented in this article is supported in the frame of the European Science Foundation EUROCORES programme *The Origin of Man, Language and Languages*, by the HFSP grant RGP 68/2002 and by the Regione Friuli—Venezia Giulia (L.R. 3/98). We are grateful to Judit Gervain for suggesting many and varied changes to our manuscript.

References

Abercrombie D 1967 Elements of general phonetics. Edinburgh University Press, Edinburgh
Baker M 2002 The atoms of language: The mind's hidden rules of grammar. Basic Books, New York
Belin P, Zatorre RJ, Lafaille P, Ahad P, Pike B 2000 Voice-selective areas in human auditory cortex. Nature 403:309–312
Bertoncini J, Bijeljac-Babic R, Jusczyk PW, Kennedy LJ, Mehler J 1988 An investigation of young infants' perceptual representations of speech sounds. J Exp Psychol Gen 117:21–33
Bertoncini J, Mehler J 1981 Syllables as units in infant perception. Infant Behav Dev 4:247–260
Bertoncini J, Morais J, Bijeljac-Babic R, McAdams S, Peretz I, Mehler J 1989 Dichotic perception and laterality in neonates. Brain Lang 37:591–605
Best CT, Hoffman H, Glanville BB 1982 Development of infant ear asymmetries for speech and music. Percept Psychophys 31:75–85
Bijeljac-Babic R, Bertoncini J, Mehler J 1993 How do four-day-old infants categorize multisyllabic utterances? Dev Psychol 29:11–721
Bloom L 1970 Language development: form and function in emerging grammars. MIT Press, Cambridge, MA
Bloom P 2000 How children learn the meaning of words. MIT Press, Cambridge, MA
Bonatti L, Peña M, Nespor M, Mehler J 2005 Linguistic constraints on statistical computations: The role of consonants and vowels in continuous speech processing. Psychol Sci 16:451–459
Bryden MP 1982 Laterality: Functional asymmetry in the intact brain. Academic Press, New York
Chomsky N 1959 A review of B.F. Skinner's Verbal Behavior. Language 35:26–58
Chomsky N 1980 Rules and representations. Columbia University Press, New York

Chomsky N 1986 Knowledge of language: Its nature, origin and use. Praeger, New York

Christophe A, Guasti MT, Nespor M, van Ooyen B 2003 Prosodic structure and syntactic acquisition: The case of the head-complement parameter. Developmental Sci 6:213–222

Christophe A, Morton J 1998 Is Dutch native English? Linguistic analysis by 2-month-olds.
Developmental Sci 1:215–219

Christophe A, Nespor M, Guasti MT, van Ooyen B 1997 Reflections on phonological bootstrapping: its role in lexical and syntactic acquisition. In: GTM Altmann (Ed.), Cognitive models
of speech processing: a special issue of language and cognitive processes. Lawrence Erlbaum,
Mahwah, NJ

Colombo J, Bundy RS 1983 Infant response to auditory familiarity and novelty. Infant Behav Dev
6:305–311

Cutler A 1994 Segmentation problems, rhythmic solutions. Lingua 92:81–104

Cutler A, Mehler J, Norris D, Segui J 1983 A language-specific comprehension strategy. Nature
304:159–160

Cutler A, Mehler J 1993 The periodicity bias. J Phonetics 21:103–108

Dauer RM 1983 Stress-timing and syllable-timing reanalysed. J Phonetics 11:51–62

Dehaene-Lambertz G, Dehaene S, Hertz-Pannier L 2002 Functional neuroimaging of speech perception in infants. Science 298:2013–2015

Dronkers NF 1996 A new brain region for coordinating speech articulation. Nature 384:159–161

Endress AD, Scholl BJ, Mehler J 2005 The role of salience in the extraction of algebraic rules. J
Exp Psychol Gen 134:406–419

Fodor J 1975 Language of thought. Crowell, Scranton, PA

Gandour J, Wong D, Lowe M, Dzemidzic M, Satthamnuwong N, Tong Y et al 2002 A cross-
linguistic fMRI study of spectral and temporal cues underlying phonological processing. J
Cogn Neurosci 14:1076–1087

Gauthier I, Skudlarski P, Gore JC, Anderson A W 2000 Expertise for cars and birds recruits brain
areas involved in face recognition. Nat Neurosci 3:191–197

Geschwind N 1970 The organization of language and the brain. Science 170:940–944

Haegeman L 1994 Introduction to government and binding theory (Blackwell textbooks in linguistics, No 1). Blackwell Publishers, Oxford

Hebb DO 1949 The organization of behaviour. John Wiley, Chichester

Hirsh-Pasek KA, Golinkoff RM 1996 The origins of grammar: evidence from early language
comprehension. MIT Press, Cambridge, MA

Jusczyk PW 1997 The discovery of spoken language. MIT Press, Cambridge, MA

Jusczyk PW, Rosner BS, Cutting JE, Foard CF, Smith LB 1977 Categorical perception of non-
speech sounds by 2-month-old infants. Percept Psychophys 21:50–54

Kanwisher N, McDermott J, Chun MM 1997 The fusiform face area: a module in human extrastriate cortex specialized for face perception. J Neurosci 17:4302–4311

Kuhl PK 1987 The special-mechanisms debate in speech research: categorization tests on animals
and infants. In S Harnad (Ed), Categorical perception: The groundwork of cognition.
Cambridge University Press, Cambridge, p 355–386

Kuhl PK, Miller JD 1975 Speech perception by the chinchilla: voiced-voiceless distinction in
alveolar plosive consonants. Science 190:69–72

Kuhl PK, Williams KA, Lacerda F, Stevens KN, Lindblom B 1992 Linguistic experience alters
phonetic perception in infants by 6 months of age. Science 255:606–608

Klima E, Bellugi U 1979 The signs of language. Harvard University Press, Cambridge, MA

Kluender KR, Lotto AJ, Holt LL, Bloedel SL 1998 Role of experience for language-specific functional mappings of vowel sounds. J Acoust Soc Am 104:3568–3582

Ladefoged P 1975 A course on phonetics. Harcourt Brace Jovanovich, New York

Landau B, Gleitman L 1985 Language and experience—evidence from a blind child. Harvard
University Press, Cambridge, MA

Lenneberg EH 1967 Biological foundations of language. Wiley, New York

Levelt WJM 1989 Speaking: from intention to articulation. MIT Press, Cambridge, MA

Liberman M, Prince A 1977 On stress and linguistic rhythm. Linguistic Inquiry 8:249–336

Manrique AMB, Signorini A 1983 Segmental durations and rhythm in Spanish. J Phonetics 11:117–128

Mazuka R 1996 Can a grammatical parameter be set before the first word? Prosodic contributions to early setting of a grammatical parameter. In JL Morgan, K Demuth (Eds.) Signal to syntax: Bootstrapping from speech to grammar in early acquisition. Lawrence Erlbaum, Mahwah, NJ, p 313–330

Mehler J 1981 The role of syllables in speech processing: Infant and adult data. Philos Trans R Soc 295:333–352

Mehler J, Jusczyk P, Lambertz G, Halsted N, Bertoncini J, Amiel-Tison C 1988 A precursor of language acquisition in young infants. Cognition 29:143–178

Mehler J, Dupoux E 1994 What infants know. Basil Blackwell, Cambridge

Meisel JM 1992 The acquisition of verb placement. Functional categories and V2 phenomena in language acquisition. Kluwer Academic Press, Dordrecht

Miller GA 1951 Language and communication. McGraw-Hill Book Company Inc, New York

Molfese D, Molfese V 1979 Hemisphere and stimulus differences as reflected in the cortical responses of newborn infants to speech stimuli. Develop Psychol 15:505–551

Moon C, Cooper RP, Fifer WP 1993 Two-day-olds prefer their native language. Infant Behav Dev 16:495–500

Morgan JL, Meier RP, Newport EL 1987 Structural packaging in the input to language learning: contributions of prosodic and morphological marking of phrases to the acquisition of language. Cognit Psychol 19:498–550

Morgan JL, Demuth K 1996 Signal to syntax: bootstrapping from speech to grammar in early acquisition. Lawrence Erlbaum, Mahwah NJ

Morgan JL, Saffran JR 1995 Emerging integration of sequential and suprasegmental information in preverbal speech segmentation. Child Dev 66:911–936

Nazzi T, Bertoncini J, Mehler J 1998 Language discrimination by newborns: toward an understanding of the role of rhythm. J Exp Psychol Hum Percept Perform 24:756–766

Nespor M 1990 On the rhythm parameter in phonology. In: IM Roca (Ed.) Logical issues in language acquisition. Foris, Dordrecht, p 157–175

Nespor M, Vogel I 1986 Prosodic phonology. Foris, Dordrecht

Nespor M, Guasti MT, Christophe A 1996 Selecting word order: the rhythmic activation principle. In: U Kleinhenz (Ed.), Interfaces in phonology. Akademie Verlag, Berlin, p 1–26

Nespor M, Mehler J, Peña M 2003 On the different role of vowels and consonants in language processing and language acquisition. Lingue e Linguaggio 221–247

Newport EL, Aslin RN 2004 Learning at a distance I. Statistical learning of non-adjacent dependencies. Cognit Psychol 48:127–162

Peña M 2002 Rôle du calcul statistique dans l'acquisition du langage (Doctoral dissertation, Ecole des Hautes Etudes de Sciences Sociales, 2002)

Peña M, Bonatti LL, Nespor M, Mehler J 2002 Signal-driven computations in speech processing. Science 298:604–607

Peña M, Maki A, Kovacic D et al 2003 Sounds and silence: an optical topography study of language recognition at birth. Proc Natl Acad Sci USA 100:11702–11705

Pike KL 1945 The intonation of American English. University of Michigan Press, Ann Arbor, MI

Pinker S 1994 The language instinct: How the mind creates language. Harper Collins, New York

Port RF, Dalby J, O'Dell M 1987 Evidence for mora timing in Japanese. J Acoust Soc Am 81:1574–1585

Premack D 1971 Language in chimpanzee? Science 172:808–822

Premack D 1986 'Gavagai!' or the future history of the animal language debate. MIT Press, Cambridge, MA

Querleu D, Renard X, Versyp F, Paris-Delrue L, Crepin G 1988 Fetal hearing. Eur J Obstet Gynecol Reprod Biol 28:191–212

Ramus F, Hauser MD, Miller C, Morris D, Mehler J 2000 Language discrimination by human newborns and by cotton-top tamarin monkeys. Science 288:349–351

Ramus F, Nespor M, Mehler J 1999 Correlates of linguistic rhythm in the speech signal. Cognition 73:265–292

Saffran JR, Newport EL, Aslin RN 1996 Word segmentation: the role of distributional cues. J Mem Lang 35:606–621

Saffran JR, Aslin RN, Newport EL 1996 Statistical learning by 8-month-old infants. Science 274:1926–1928

Segalowitz SJ, Chapman JS 1980 Cerebral asymmetry for speech in neonates: a behavioral measure. Brain Lang 9:281–88

Seidenberg MS, MacDonald MC 1999 A probabilistic constraints approach to language acquisition and processing. Cognit Sci 23:569–588

Selkirk E 1984 Phonology and syntax: the relation between sound and structure. MIT Press, Cambridge, MA

Stevens KN 1998 Acoustic phonetics. MIT Press, Cambridge, MA

Villringer A, Chance B 1997 Non-invasive optical spectroscopy and imaging of human brain function. Trends Neurosci 20:435–442

Wanner E, Gleitman LR 1982 Language acquisition: the state of the art. Cambridge University Press, Cambridge

Werker JF, Tees RC 1983 Developmental changes across childhood in the perception of nonnative speech sounds. Can J Psychol 37:278–286

Werker JF, Tees RC 1984 Phonemic and phonetic factors in adult cross-language speech perception. J Acoust Soc Am 75:1866–1878

Yang C 2004 Universal grammar, statistics or both? Trends Cog Sci 8:451–456

DISCUSSION

Logothetis: What you have been describing can be mathematically represented as a series of Markov chains. I thought the recent results from Mark Hauser with tamarin monkeys give a bit of a hint as to what might be happening. They have been testing the ability of these animals to detect certain sequences, randomising the process appropriately with Markov chains. They have shown that the tamarins have the basic machinery for detecting certain transitions. All is needed is for them to fine tune it. Would this be a mechanism that appeals to you?

Mehler: What Fitch & Hauser defined were 'A' items and 'B' items. Both As and Bs were syllables. There were eight 'A' syllables and eight 'B' syllables. Using these 16 syllables they tried to teach two types of grammar to the tamarins. One is an $(AB)^d$ kind of grammar while the other is an $A^n B^n$ grammar. While the first generates sequences of syllables like (ABAB, ABABAB, etc.), the other generates sequences like (AABB, AAABBB, etc.). Unfortunately, one has to explain to the animal which syllables are As and which are Bs. They signalled As to the animals using a high pitched voice and the Bs using a low pitched voice. Their results suggest that while humans extract both kinds of grammars from the few examples they are given, the tamarins only extract the simpler grammar, namely, the $(AB)^n$. Perhaps the monkeys aren't paying attention to the syllables at all but only to the high and low pitch tones. They may even be computing the transition probabilities between high-pitched and low-pitched sounds. I believe that their results can be explained in this way. Of course, some further tests could be used to explore whether both populations learned the intended grammar or something else. I will not reveal at this point how adults in our lab behaved when they are given sequences like HLHLH or HHHLL neither of which is grammatically compatible with the grammars that the authors tried to teach the tamarins and their human participants. However, our own work suggests that the convergence on a grammar required more than Fitch and Hauser think. Another question is whether humans but not tamarins learn some kinds of grammars and not others. Obviously this must be true, otherwise tamarins would by now be using fully-fledged grammatical structures. In short, Fitch and Hauser haven't demonstrated that the tamarins are learning either of these grammars and even the behaviour of humans needs to be evaluated with more telling tests.

Haggard: At the end of your talk it seems you were suggesting that the consonant plays a special role as the carrier of the linguistic unit. It is interesting, but I'm not sure I understand why the consonant does this and not the vowel. In particular, could it ever be the other way round?

Mehler: We explored the phonologies of many languages spoken throughout the world. Languages with lexical roots that are characterized by vowels were not found while languages in which the sequence of consonants defines lexical roots are numerous. Indeed, in Semitic languages a sequence like *GDL* is not a word but a root that can realize itself as: *gadol*, 'big' (masculine adjective); *gdola*, 'big' (feminine adjective); *giddel*, 'he grew' (transitive verb); *gadal*, 'he grew' (intransitive verb); *higdil*, 'he magnified' (transitive verb); *magdelet*, 'magnifier' (lens), etc. Thus, *GDL* is a root whose meaning is related to the 'enlarging/growing' semantic complex. Why, we can now ask, is it not possible to find similar roots defined in terms of the sequence of vowels?

Marina Nespor and colleagues have written a paper (Nespor et al 2003) entitled 'On the different role of vowels and consonants in speech processing and language acquisition', suggesting the following: there is not much you can do with a consonant except to pronounce it or mispronounce it, while with vowels the speaker can do a lot of things among which stressing it, changing slightly its volume or the typical first and second formants to suggest a given dialectal variation, etc. In other words, we have reasons to believe that to a large extent vowels tend to influence grammatical properties, while consonant are mostly related to the characterization of the lexical items in a language

Rumiati: Another example for me would be the *r* for me or the *r* for the Hebrew speaker. They are simply characterizing the fact that we are speaking different languages which have different phonological features.

Mehler: Even I, who have a terrible accent in every language I speak, am happy to notice that after a while my listeners tend to understand and comprehend what I say. Thus, mispronunciations tend to be located mostly in my vowels and if they also attain the consonants it must be in a systematic fashion, like in my fricatives or nasals. Yet, vowel mispronunciations have been used since ancient times to identify foreigners: the word 'shibboleth' was used to discriminate foreigners in the Bible. But as far as I know, in that example, the variation is at the initial s/sh, not in the vowels!

Logothetis: So does the fact that the Hebrew script can be written without vowels mean that it gives you much more range of expressing the vowels?

Barash: There is an interesting point here. Hebrew is extreme in the sense that there is an Ashkenazi pronunciation which was used by the Jews in Europe and the other pronunciations which were used by Jews elsewhere. They are very different. For me it is difficult for me to understand Ashkenazi Hebrew.

Diamond: I have a question about the learning of grammatical rules and the distinction between A^nB^n and (AB^n). You said that ABAB can't be (AB^n), but does it really violate the rule?

Mehler: The actual grammar that can relate to Chomsky is not strictly speaking A^nB^n: Rather it describes nested syntactic constructions as we often use in natural languages. Consider, such constructions as

The boy that my mother's cousin met yesterday fell down the stairs

Clearly 'the boy' is linked to the phrase 'fell down the stairs' while 'the mother of my cousin' did not fall down the stairs. Rather 'mother' and 'cousin' met yesterday. We link 'the boy' to the 'fall down the stairs' and the phrase 'my mother's cousin met yesterday' is another constituent that is located between two parts, one at each of the extremes of the sentence. It is these nested constructions that are being referred to. If you leave out one of the constituent phrases the sentence as a whole becomes uninterpretable. None of these properties were really tested by Fitch & Hauser when they claim that humans learn an A^nB^n grammar.

Dehaene: Can you speculate on the relationship between perceptual development in the language domain and the development of the speech production system? Is it possible that there is, very early on, a covert, internal mapping between the perception and action systems? One bit of data is that when we do functional imaging in infants at three months of age, we can now reproducibly show that Broca's area is already activated when the infants are listening to their maternal language.

Mehler: That is a critical question. Twenty-five years ago we proposed that the syllable was incomprehensible if not the 'atom' of perception. Independently, Levelt (1989) showed that the syllable also acts as an atom of production. It would be strange if the perceptual procedures carry us to representations that are neither similar nor connected to the representations used to generate speech acts. Do we have direct evidence that this is so? No. Much more evidence is needed. It is exciting to notice that methods have become available to do these kinds of experiments with very young infants. Let me illustrate this with a very simple experiment. As soon as we know that an infant has learned a dozen or so words, is it possible to show that whenever the baby listens to one of those words areas that are also active during production become activated? And do such areas become more activated when the infant listens to nonce words, i.e. to *detty* instead of *teddy* there would be less activation than when the infant listens to *mimmo* than to *mommy* as the aforementioned hypothesis should predict.

Derdikman: Do you have a suggestion for why there is such a relationship between the phonetic structure of a language and its syntax?

Mehler: We have thought a lot about this question over the past ten years. First we noticed, as predicted by Nespor & Vogel (1986) that there must be some relation between prosodic aspects and syntax. The first evidence suggesting that the conjecture might be correct was when we discovered that very young infants can discriminate utterances drawn from two different pairs of languages but not from any two pairs of languages (Mehler et al 1988, Nazzi et al 1998, Ramus et al 1999). From these studies, Ramus et al (1999) proposed that rhythm (as measured by the quantity of vowel time in the typical utterance of the language and the variability of the intervocalic intervals) predicts the infants' behaviour. More recently yet, we showed that if one plots 20 highly varied languages in a rhythmic chart there is a dividing line that separates the Head-Complement languages (as most Romance languages) from Complement-Head languages (as Japanese and Basque) (Mehler et al 2004). Research in progress suggests that there may be a much more intimate relation between the sound structure of languages and the syntax they implement.

References

Gergely G, Bekkering H, Kiraly I 2002 Rational imitation in preverbal infants. Nature 415:755

Levelt WJM 1989 Speaking: from intention to articulation. MIT Press, Cambridge, MA

Mehler J, Jusczyk P, Lambertz G, Halsted N, Bertoncini J, Amiel-Tison C 1988 A precursor of language acquisition in young infants. Cognition 29:143–178

Mehler J, Gallés NS, Nespor M 2004 Biological foundations of language: language acquisition, cues for parameter setting and the bilingual infant. In: Gazzaniga M (Ed) The new cognitive neurosciences III. MIT Press, Cambridge, MA

Nazzi T, Bertoncini J, Mehler J 1998 Language discrimination by newborns: towards an under-
    standing of the role of rhythm. J Exp Psychol Hum Percept Perform 24:756–766
Nespor M, Vogel I 1986 Prosodic Phonology. Foris, Dordrecht
Ramus F, Nespor M, Mehler J 1999 Correlates of the linguistic rhythm in the speech signal.
    Cognition 73:265–292

Final general discussion

Diamond: I notice that several speakers referred to some of the pioneers of psychology, neuroscience and experimental psychology in their papers, in referring to the inseparability between percepts and actions. People such as Sherrington and Adrian have made observations of this sort. I noticed the reactions of the audience: people seemed to agree with this idea of non-separability. Yet when Nikos Logothetis asked us to try to begin to define the decision-making network, I was surprised that most people agreed that we should drop the sensory part of it—at least what some people referred to as a purely sensory part—and then consider what is left in the network to be the decision-making part of it. This contradicts the agreement with the initial proposal of inseparability between percepts and actions. In the end it may be a useless exercise to try to define the transition between sensations and actions. Nevertheless, I want to reopen that question with a thought experiment. Suppose that a stimulus *a* produces a percept also called *a*, and we ask people to give a reaction *a′* when they experience this. Stimulus *b* produces a reaction *b′*. Suppose that we can change the subject's reaction through an external device. Does coming to the opposite action affect their judgement of the stimulus that occurred before? I wouldn't be too surprised if how you react to a stimulus affects your interpretation even though the stimulus has occurred before. For example, if we see a face talking, the visual input is so salient that we are convinced that the voice comes from the face. If the voice comes from a different source we continue to attribute the voice to the speaking face, and so we reinterpret the time of sound arrival to our ears according to a decision that we have made. Decisions thus affect percepts. Should we exclude sensations from the decision-making process, or is there a seamless transition?

Rizzolatti: I think it is important to keep sensation and perception separated. Think, for example, of the McGurk effect. Individuals are presented with two syllables ('ba', 'da', 'ga') simultaneously, one in the auditory and in the other in the visual modality. When the syllable presented in one modality does not match the one presented in the other modality, the individual may perceive a syllable different from both those presented. There is no reason to doubt that both visual and auditory stimuli are correctly analysed (that is the sensation is correct), yet the percept is different. When I say that perception and action results from a common substrate, I am not talking about what happens in the retina or in the cortical representation of the whiskers. In the syllable case, what is perceived depends on the language motor areas.

Sensation is a distinct process from action, although it may be influenced by it. In contrast, perception and action share the same neural substrate.

Haggard: The executive areas of the brain structure determine the incoming afferent sensation. Even primary cortex can be preset by executive areas and multimodal areas to process stimuli in a particular way. This seems not that different in principal from the active touch idea that you control your own sensory input by movement. Except that in the case of executive control of unimodal areas you aren't using your body to control your input, you are doing it entirely internally in your brain. If you wanted to be radical, you could say that in both of these cases the cognitive brain is setting up the afferent transmission to acquire good, better, optimal information. We were talking about where decisions are made. David Sparks very nicely said that it must be made before the relevant neurons in the superior colliculus fire. I think it must be made *after* active touch: if I am carrying out active touch, or my frontal lobes are preparing my somatosensory cortex for some input, then by definition I haven't yet decided what the stimulus is, and I am still trying to improve the sensory information I have about the stimulus. That's what this descending signal means.

Scott: There is an illusion generated in a rotating room in Jim Lackners' lab. First, you stand at the side of the room and get used to the velocity of the rotating room. When you make your first movement directly in front of you your arm gets 'knocked' to the side and you feel this imaginary force on your arm. Within a few movements you move straight and no longer feel any force. If the room rotation is stopped you get 'knocked' in the opposite direction. Within a few movements it is gone. This percept is completely generated from actions and what you are expecting from the sensory periphery.

Logothetis: Your perception under these conditions is also affected. I have been in that room, and the angles do not appear to be 90 degrees any more. It is not just the motion that is changed.

Scott: The sensation of the apparent force on your limb is changing. You adjust in just a few movements, and this is only your arm movements that have created that.

Treves: I was just thinking of a class of experiments done by Edmund Rolls in which he used gustatory stimuli and fed subjects to satiety. In primary cortex there is selectivity that is not affected by satiety, but in secondary cortex there is satiety. There is a gradient along the sensory cortex of how much something that is not in the stimulus can affect things.

Diamond: We would expect those gradients to be different for different systems, animals and paradigms. It would be interesting to explore them for each perceptual experience.

Sparks: When I began my career we knew a lot more about sensory neurophysiology than we did about motor neurophysiology, certainly at a cellular level. We could do the sensory neurophysiology on anaesthetized animals, but it is hard to record single-cell activity related to movements in a paralysed animal. The motor physiologists therefore lagged behind the sensory physiologists in terms of cellular understanding. Because of this the sensory physiology has dominated the way we study sensory systems. The point I want to make is that in sensory systems, perception and cognition are not the only endpoint of sensory processing. Neither sensation nor cognition has any adaptive value in the absence of action. The brain has evolved to translate these sensory signals into motor commands. The motor system imposes constraints on the types of sensory processing that must occur. The normal way we do sensory neurophysiology ignores all of that. We should look at the types of signal transfer mechanisms that are required to interface with the format of the motor command, and new areas of sensory neurophysiology will open up. In terms of motor physiology, I'll stick to eye movement. It is well known that the execution of an eye movement is influenced by cognitive factors. There are some things that aren't typically measured that might be more sensitive than just measuring probability or latency of movement. These are the speed and duration of the movement. If you are doing neurophysiological recordings and you have neural activity that is cognitively mediated that you think may be influencing the execution of the movement, there is an optimal time to measure it. The thing to do is remember that the saccadic system is a gated system. It is only when the omnipause neurons (OPNs) are turned off that the commands to produce a saccade can occur. The activity that is going to influence the execution phase of the movement is the activity that is present at the time the movement is executed.

Also, the saccadic system has properties that can be used to assess the presence and magnitude of cognitive influences. One is that the superior colliculus has a map and its retinal and auditory inputs can activate different parts of the map simultaneously. If this occurs, the system does a vector average. It is possible to demonstrate the presence of a cognitive input using this feature of the motor circuitry. Gold and colleagues have done experiments in which a region of the brain that produces a saccadic eye movement was stimulated and looked at the development of cognitive influences by studying the trajectory of the movement. As the signal increases the stimulation-evoked movement will deviate from the control trajectory to an intermediate trajectory. This will build up in time. This is a sensitive way to assay the presence and magnitude of these cognitive variables. If you present a noise burst and look at an acoustically induced saccade, often they have a curved trajectory. Van Opstal and colleagues suggested that this was because the azimuth and elevation cues are quite different (Frens & Van Opstal 1995). It is the time and intensity of interaural differences that code information about the azimuth, but it is the spectral cues dependent on high frequency input that give elevation cues. They speculated that there is a delayed vertical component because processing the spectral cues takes longer. When they manipulated the frequency of the noise burst they could vary the amplitude of the vertical component. These mapping properties can be used as sensitive measures of the presence and amplitude of cognitive influences.

Gold: The idea was using the effect of this vector average as an assay. The microstimulation produces an eye movement of a known vector. If it evokes an intermediate trajectory this can be used as an assay of the other activity, which is what we think of as this transformed sensory variable into motor coordinates.

Schall: There are two points I would like to make. First, it seems that this word 'decision' is being used frequently, perhaps carelessly, and out of context. Sometimes monkeys and people are faced with alternatives and choose between them for the purposes of achieving a goal. The word 'choice' can explain this kind of behaviour. I believe the word 'decision' should be reserved for those cases when there is real deliberation and the consequences are higher and more ambiguous. This is what competent humans do and are held responsible for. I think it is fair for us to ask whether macaque monkeys in physiology laboratories are ever deliberate? Even when the random dots produce 2% motion strength, are the monkeys deliberating? Perhaps not. We are certainly studying processes related to choice behaviour, but we need to be careful before we say that this is how decisions are made. My second point is that the title of this meeting is 'Percept, decision and action'. The claim is that there is nothing in the middle. There is sensation, the brain sorts it out, and then there's the mapping to the action. The complexity of our behaviour comes from our ability to map arbitrary responses onto given stimuli, but it is this mapping where all the action is taking place. Looking for a discrete decision stage distinct from the sensory representation and the motor preparation may be a fool's errand.

Derdikman: Related to your last comment, I believe that we make the mistake of assigning a decision process where it is not appropriate because we are so familiar with our own language. We have the term 'decision'. Every time we make a decision we can also be thinking of ourselves knowing that we are making a decision. We are very reflective about the things we do. However, monkeys are much less reflective. It could be that we are actually trying to impose the term 'decision' that is so familiar to us on the other species, where there is perhaps no such thing as decision making in the sense we use it as human beings.

Albright: Is accumulating information different from deliberating? We know that if it takes more time we are accumulating information to make the choice. Is that qualitatively different from what you are calling 'decision'?

Schall: I will claim that it is. It is possible to choose in the sense of acting in the context of alternatives, even when they are vague, in a more automatic sense, for example ordering a meal at a favourite restaurant. But deliberating about complex decisions, like ordering a meal at an unusual restaurant, cannot be done while you are doing something else. Deliberation entails other cognitive processes such as working memory that we know requires dedicated resources.

Albright: You do accumulate information, though. You read through the things that are on the menu; you build up something to base your choice on.

Schall: That is a natural way to think of it, but we are not guaranteed that this is the mechanism that holds for decisions such as those made by political leaders contemplating war, for example.

Barash: My bias is to think that there are intermediate states between visual and motor responses in eye movement. With regard to decisions, there are choices that monkeys make that are more automatic and less automatic.

Hasson: Humans, more than any other primate, are first and foremost social creatures. Therefore our decision to perform a certain act should be appropriate to each given social context. Perhaps the mirror system is a prime example of a system that was designed for shaping our social behaviours. This system is designed to adjust our behaviour by learning from what other people are doing. Moreover, as Rizzolatti showed in his talk, this system is highly sensitive to contextual cues. So one can conceive of the mirror system as a decision making device that intends to directly link our perception to our actions in this world.

Harris: Perhaps a way to think about a distinction between choice and decision is that the choice is focused on the stimuli in the external world, whereas a decision involves reflecting on your own action and the consequences of it.

Wolpert: We're just going to make this into a discussion of consciousness, and get stuck there. It seems like we are almost making decisions into a conscious internal discussion, and we'll then be stuck with all the same problems.

Schall: But that is the decision-making people care most about. It may be that the mechanisms of the brain are the same when we decide who to marry as when we select a soft-drink from a vending machine, but we cannot assume this.

Wolpert: If we are looking for the neural correlate of decision making, we know that one consequence of a decision is a motor act, so it is very hard to dissociate a motor act from neural correlates of decision. How are you going to do this neurophysiologically?

Schall: I decided to come to this meeting months ago, but I didn't come until two days ago. Clearly, we can choose in advance. Monkeys can too, although they probably cannot plan in advance in a manner as complicated as we.

Wolpert: Squirrels hide nuts: is this planning in advance? But you don't believe that they have thought about this.

Ditterich: Doesn't it happen relatively often that we are facing a difficult problem when we say that we will sleep on it? Then we wake up and we have made up our minds. What has happened? Was it some kind of automatic process or were you deliberating what you should do in that situation?

Schall: We say we make or take decisions, but we don't really. If it is really complicated, we say we can't make up our mind. We don't have access to how we make our decisions. They happen to us.

Diamond: That's true for every brain process, isn't it?

Wolpert: Ben Libet says we just get informed of our decisions after the motor system has made them.

Haggard: These were interesting studies in the 1980s which other people includ-ing myself have followed up (Libet et al 1983, Haggard & Eimer 1999). These studies are concerned with concepts of voluntary action independent of a stimu-lus. If you ask people when they first experienced the intention to make an action, on average they experience intentions a couple of hundred milliseconds before action. But of course the brain has begun to prepare the frontal motor potentials around a second before. There is a long period where your brain knows you are going to move when you don't. Philosophers love this. We need to distinguish care-fully, though, between the concept of internally generated actions, which are only very remotely connected to a specific stimulus, and the sorts of situations which are more in focus at this meeting, where there is a set of stimuli and a set of responses, and perhaps a rather open relationship between them. I'm inclined to agree with Jeff Schall: where we are thinking about a mapping and can see a clear feed-forward link between stimulus and response, then decision and deliberation are perhaps not the way to think about it. We want to think more about perceptual categorization and feature extraction. At what point do we move from just mapping into real decisions? The words that seem to me to be relevant are things like induc-tion, hypothesis making, somehow going beyond the information that is immedi-ately present in the stimulus. How do we do this? The work of Jerry Fodor is probably relevant here. He has done some important philosophy of cognitive science. He envisaged the input and output sections of the human mind as being modular, feedforward. Then there is a non-modular central soup in the middle. He claimed that these work en bloc. They formed a very general representation of a whole series of beliefs all of which will influence the gap between the output of the sensory modules and the start of the motor modules. His view was that this Quinean inter-related property of our central representations makes them intractable to science. I am confident now that this is wrong. For example, neuro-scientific experiments on the brain basis of context effects show us how these inter-mediate stages operate.

Rizzolatti: I think there are a lot of data from neuroscience, from the classical work of Mouncastle on the parietal lobe (e.g. Mountcastle et al 1975) to the dis-covery of mirror neurons—that proves that Fodor is wrong: the mysterious some-thing between sensory and motor doesn't exist. I think we have to make realistic, neurophysiological hypotheses on higher-order cognitive processes, not to think of them as something not amenable to the scientific enquiry.

Haggard: Fodor gave up at the central point and said that we can't be scientists here. But I think we can and we should, even in the frontal cortex.

Treves: I like your characterization of complex decisions as deliberations. I would suggest a kind of Alan Turing view: maybe we should not criminalize monkeys for making simple decisions, but in human decisions, the things we are *not* so interested in are precisely those which we *can* describe. The Turing idea would be that you can provide a mathematical description of a phenomenon, like the beautiful description from Daniel Wolpert's talk. This is not what we would call a deliberation. A deliberation is something that we have difficulty describing mathematically. These are the challenges that we should address: to develop descriptions of mathematically intractable *deliberation-making*.

Diamond: Almost every sensory cortical area projects through layer 5 into motor centres. Sensory cortical areas can have a direct influence on complicated decisions.

Krubitzer: They also have very strong projections to the thalamus, which we haven't discussed. The thalamus is a very quick and powerful way of modulating incoming sensory stimuli from S1 through different levels of the cortex. The thalamus has massive input from the cortex as well as from sensory receptors. Psychophysical experiments show that detection levels are modified rapidly by what occurred prior to that. We could simply be modifying the ratio of sensory inputs coming in through the thalamus. We talk about decisions as unitary phenomena when in reality they might not be.

Derdikman: Two comments. First, I have one for Jeff Schall. Think of a huge crowd in the arena at ancient Rome: who makes the decision about whether to kill the fallen gladiator or not? It seems improbable to assume that there was a single person who was making the decision about the fate of the gladiator. Second, for Mathew Diamond, an experiment comes to mind. It is an old experiment by Held and Hein. Two kittens are sitting in two baskets. The first can walk, while the second is moved by the first kitten. Their perception of the world is totally different, although both of them have had exactly the same sensory experience.

Dehaene: As humans, we have a fairly clear, introspection of when we engage in conscious decision or deliberation, and when we do not. To track decision making in the human brain, one suggestion would be to capitalize on this distinction which is available to humans. It seems to me that at this meeting, we would have benefited from a closer examination of the neuropsychological literature, which is very clear in some respects. Consider for instance the 'alien hand' syndrome: some patients declare 'my hand is moving, but I am not in command of that action'. There are also other paradigms that allow examination of this distinction in normal subjects. I am reminded of a simple experiment by Marcus Raichle which examined the neural bases of automatization of behaviour. If you are asked to generate a verb in response to a noun, the first time you do this you have to go through a process of deliberation. You are searching for the appropriate verb for that noun. If you do that 10 times with the same list of nouns, however, then you automatize the process of associating verbs to nouns. It is possible to image the brain activation contrasting these two states. Many areas have the same level of activity, but the parieto-fronto-cingulate network changes drastically with activity reducing when the process is automatized. It seems to me that the bulk of evidence points to a crucial role of long-distance parieto-fronto-cingulate networks in conscious decision making, as stressed by my colleagues and I in the 'global neuronal workspace' model (Dehaene & Naccache 2001, Dehaene & Changeux 2004).

References

Dehaene S, Changeux JP 2004 Neural mechanisms for access to consciousness. In M. Gazzaniga (Ed.) The cognitive neurosciences, 3rd edition. Norton, New York, Vol 82, p 1145–1157

Dehaene S, Naccache L 2001 Towards a cognitive neuroscience of consciousness: Basic evidence and a workspace framework. Cognition 79:1–37

Frens MA, Van Opstal AJ 1995 A quantitative study of auditory-evoked saccadic eye movements in two dimensions. Exp Brain Res 107:102–117

Haggard P, Eimer M 1999 On the relation between brain potentials and the awareness of voluntary movements. Exp Brain Res 126:128–133

Libet B, Gleason CA, Wright EW, Pearl DK 1983 Time of conscious intention to act in relation to onset of cerebral activity (readiness-potential). The unconscious initiation of a freely voluntary act. Brain 106:623–642

Mountcastle VB, Lynch JC, Georgopoulos A, Sakata H, Acuna C 1975 Posterior parietal association cortex of the monkey: command functions for operations within extrapersonal space. J Neurophysiol 38:871–908

Sigman M, Dehaene S 2005 Parsing a cognitive task: a characterization of the mind's bottleneck. PLoS: Biology 3:e37

Index of contributors

Subject index